Analytical Groundwater Modeling

Flow and Contaminant Migration

WILLIAM C. WALTON

Library of Congress Cataloging-in-Publication Data

Walton, William Clarence.
  Analytical groundwater modeling
        p.   cm.
  Bibliography: p.
  Includes index.
  1. Groundwater flow — Computer programs.   I. Title.
GB1197.7.W336   1988   551.49 — dc 19   88–23831
ISBN 0-87371-178-5

Third Printing 1991

Second Printing 1989

LEWIS PUBLISHERS, INC.
121 South Main Street, Chelsea, Michigan 48118

PRINTED IN THE UNITED STATES OF AMERICA

Preface

Four analytical microcomputer programs for quick and easy simulation and graphing of uncomplicated two-dimensional groundwater flow and contaminant migration situations are presented in this book. Their operation is explained, and program concepts, techniques, and methods are described. These programs are designed for use with inexpensive hardware, and complement numerical techniques which should be used in complicated situations where deviations from idealizations assumed in analytical models are significant. Target readers are practicing engineers, geologists, and hydrologists; water well contractors; educators and students; governmental personnel; industry representatives; and others. The reader is expected to have a working knowledge of hydrogeology, be acquainted with the BASIC language, have microcomputer operating experience, and have read books on the hydraulics of groundwater such as that written by Bear (1979).

The BASIC programs feature approximate simulations of the following aquifer, sink, and source conditions: nonleaky artesian, leaky artesian, water table, barrier and recharge boundaries and discontinuities; production and injection wells, drains, and mines; and contaminant slug and continuous source areas and plumes, respectively. Medium quality screen graphic subprograms interpolate surface data and display drawdown and contaminant concentration arithmetic and semilog temporal graphs and spacial contour maps. Program simplicity makes the program listings easy to follow and to alter.

The programs WELFUN, WELFLO, CONMIG, and GWGRAF are tailored to the IBM PC (IBM Personal Computer, trademark of International Business Machines Corporation) and compatible microcomputers. They are prompt/option driven, contain no menus, and feature user-friendly interactive data base entry and frequent screen displays of helpful information and instructions. Aquifer, source, and sink situations are user-defined by entering an appropriate letter designated for desired options in response to prompt message lines. Input/output options include: data base hard copies produced by a printer and data base diskette files; spacial and temporal drawdown or recovery and contaminant concentration diskette files, screen displays, and hard copies; and diskette output files for export to commercial graphics programs.

The simplistic structure of the programs, although not especially efficient, should enable the reader to modify and extend the programs as desired with little difficulty. It should be emphasized that the programs in this book are not fancy but plain. They are useful only within the context of the assumptions and simplifications on which they are based. The diskette software provided with the book is not meant to be commercial and is intentionally crude. The user may wish to add a menu and means by which a data base may be imported after data manipulation with commercial data base or spreadsheet software.

Emphasis has been placed in this book on practice rather than theory. Simulation equations are presented in abbreviated and final form. Mathematical derivations of equations may be pursued through references. Subject matter is presented in as nonmathematical a manner as possible.

I hope that this publication will assist the reader in understanding and appreciating the reasoning behind popular analytical groundwater model and graphics microcomputer program algorithms, and thus enhance the fuller use of available software. I have directed the reader's attention to literature concerning more advanced analytical models than those presented herein.

The book is based on my lecture notes for the International Ground Water Modeling Center short course, "Practical Analysis of Well Hydraulics and Aquifer Pollution," held at the Holcomb Institute, Butler University during the early 1980s and the "Groundwater Concepts and Modeling" short course held at Boise State University on February 26-28, 1986. The comments and reactions of around 125 students attending the lectures were most helpful in the formulation of this publication.

William C. Walton

Mahomet, Illinois

William C. Walton is a semiretired Consultant in Water Resources who resides on the banks of the Sangamon River at Mahomet, Illinois. Bill received his BS in Civil Engineering from Lawrence Institute of Technology in 1948 and attended Indiana University, the University of Wisconsin, Ohio State University, and Boise State University. He served for 10 years as Director of the Water Resources Research Center and Professor of Geology and Geophysics at the University of Minnesota.

Bill's 40 years of experience in the water resources field include 1 year with the U. S. Bureau of Reclamation, 7 years with the U.S. Geological Survey, 6 years with the Illinois State Water Survey, and 9 years with consulting companies such as Camp Dresser & McKee Inc. and Geraghty & Miller. He was Executive Director of the Upper Mississippi River Basin Commission for 5 years and has participated in water resources projects throughout the United States and Canada, and in Haiti, El Salvador, Libya, and Saudi Arabia.

The positions he has held include: Vice President of the National Water Well Association; editor of the journal

Ground Water; Chairman of the Ground Water Committee of the Hydraulics Division, American Society of Civil Engineers; member of the U.S. Geological Survey Advisory Committee on Water Data for Public Use; Consultant to the Office of Science and Technology, Washington, D.C.; and Advisor to the United States Delegation to the Coordinating Council of the International Hydrological Decade of UNESCO.

Bill also served as a Visiting Scientist for the American Geophysical Union and the American Geological Institute and has lectured at many universities throughout the United States. He is the author of 5 books and over 70 technical papers.

Contents

List of Figures and Tables

Figures

Tables

1

Introduction

There are many analytical models which may be used to simulate groundwater flow and contaminant migration (see Walton, 1988, pp. 141–335). Analytical models are utilized in the analysis of pumping test data, in predictive aquifer performance and contaminant transport with uncomplicated aquifer conditions, in preliminary estimates of complicated well and contaminant source impacts, and in numerical model design and verification. Numerical models (see Kinzelbach, 1986) are more realistic and adaptable than analytical models and are generally applied to complicated situations. However, models should be in tune with the data base and the scale of decisions to be affected by model results. In many cases, basic data and the scale of decisions are not sufficient to warrant a rigorous description of complex aquifer conditions, and analytical models may be the most appropriate, efficient, and economical modeling approach.

Sophisticated graphics commercial software packages with high presentation quality but intensive user effort are available for creating drawdown and contaminant concentration temporal and spatial XY graphs, maps, and 3-D plots. However, medium quality graphics created on the screen and sent to a printer with the simple but crude programs presented herein require little user effort and often are the most useful alternative.

This book presents four microcomputer programs, WELFUN, WELFLO, CONMIG, and GWGRAF, to foster greater use of analytical groundwater models and simple graphics programs in appropriate situations. The programs are written in IBM PC BASICA. They may also be run on Microsoft GW-BASIC (trademark of Microsoft Corp.) on IBM-compatible microcomputers that run on the MS DOS operating system, or converted for use on nonIBM-compatible microcomputers (see Harris and Scofield, 1983). Source codes for the programs are listed in Appendixes A–D. Two diskettes containing complete sets of the BASICA programs and executable files produced with the QuickBASIC version 4.0 (trademark of Microsoft Corp.) Compiler are included with this book. Complete instructions on how to run the BASICA and executable programs are presented in Appendix E. Abbreviations used in this book are: feet (ft), seconds (sec), minutes (min), gallons per minute (gpm), cubic feet per second (cfs), gallons per day per foot (gpd/ft), gallons per day per square foot (gpd/sq ft), meter (m), milligrams (mg), and liter (L).

Programs WELFUN, WELFLO, and CONMIG calculate common well function values and simulate a wide range of groundwater flow and contaminant migration situations including: drawdown or recovery due to multiple production and/or injection wells with variable discharge or recharge rates, drains, and mines under nonleaky artesian, leaky artesian, and water table conditions with barrier and/or recharge boundaries and discontinuities; and development of localized contaminant plumes from slug or continuous source areas of various shapes and sizes with advection, dispersion, adsorption, and radioactive decay.

GWGRAF creates medium quality graphs with convenient automatic preselection of complex graph features. Arithmetic XY graphs of time-drawdown or time-concentration, semilog XY graphs of time-drawdown, and drawdown or contaminant concentration contour maps may be created with program GWGRAF. Also, drawdown or contaminant concentration surface triangulation interpolation may be accomplished with GWGRAF.

Programs will give reasonable answers, providing the

user-defined data base is appropriate. The reasonableness of data base values may be ascertained with information presented in Appendix F. Selected data bases and associated exact well function, drawdown, and concentration values are presented in Appendix G for user verification of program operation. Example input/output displays are presented in Appendix H to assist the user through initial program operation.

Execution times for most WELFUN, WELFLO, CONMIG, and GWGRAF BASICA subprograms will be a few minutes. Execution times may be shortened (speed-up factors of 3 to 20) by compiling programs with available software such as the QuickBASIC compiler version 4.0 or later (trademark of Microsoft Corp.), Microsoft BASIC Compiler version 6.0 or later, or TURBO BASIC (trademark of Borland International, Inc.) which allow numerical processing in the 8087/80287 math coprocessor environment. Graphics programs may be speeded up by translating QuickBASIC to QUICKC with the program B-TRAN (trademark Software Translations, Inc., Newburyport, MA).

Integrated scientific/engineering personal computer graphics packages are available for creating XY graphs, contour maps, and 3-D plots with much higher resolution than can be obtained with GWGRAF. The user should contact the International Ground Water Modeling Center, Holcomb Research Institute, Butler University, Indianapolis, IN (IGWMC) for a list of microcomputer graphic software for the groundwater industry.

For example, GRAPHER (trademark of Golden Software, Inc., P.O. Box 281, Golden, CO 80402) displays black and white or color linear-linear, log-linear, linear-log, and log-log XY graphs. It supports the IBM PC, XT, or AT (or compatibles) and a wide variety of printers and plotters. GRAPHER has five different types of curve fitting to control points: linear, logarithmic, exponential, power, and cubic spline. Grid lines may be solid or dashed and multiple scales may be displayed on a single graph. Axes may be positioned at any location and axis tick marks and labels are user-defined. Lines through data points may be solid or dashed. Other features include: user-defined data point symbols, data file (free ASCII) import or keyboard spreadsheet data entry, data point labels, superposition of graphs, zoom and panning, text selected from user-defined symbols sets, and text rotated to any angle and scaled to any size.

SURFER (trademark of Golden Software, Inc.) creates contour maps and 3-D surfaces. It imports ASCII data files or creates a spreadsheet data base from the keyboard. Inverse distance or kriging techniques are available for interpolation between irregularly spaced data points. (Kriging is a method for optimizing the estimation of a magnitude, which is distributed in space and is measured at a network of points.) SURFER displays black and white or color contour maps with: in-line contour labels scaled to any size and user-defined symbol sets; bold, dashed, or normal contour lines; hachure marks; smoothed or unsmoothed contour lines; irregular contour intervals; user-defined scaling and sizing; location posted by user-defined symbol sets and labeled with numeric or text strings; axes with tick marks and labels on any side of the map; and multiple shaped boundaries. Black and white or color 3-D surface plots may be created with: stacked contours or fishnet surfaces; perspective or orthographic projections; hidden line removal; rotation to any angle; smoothed or unsmoothed lines; irregular stacked contour intervals; user-defined scaling and sizing; locations posted by any symbol and labeled with numeric or text strings; 3-D axes with tick marks and labels; multiple shaped boundaries; and the top, bottom, or both, of the surface. A surface plot from a user-defined function may be displayed. SURFER supports IBM PC, XT, or AT (or compatibles) and a wide variety of printers and plotters.

GRAPHER and SURFER have an "import" option to load (X,Y,Z) or (X,Y) data from ASCII (text) files generated under other programs or created with a word processor or an editor such as IBM's EDLIN editor on the PC-DOS (trademark of International Business Machines Corp.) diskette. Files with a .DAT filename extension must contain one (X,Y,Z) or (X,Y) entry per line; values on a single line must be separated by either a space, tab, or comma. WELFLO and CONMIG contain an option to create a file with a format matching that required by SURFER.

The program OMNIPLOT (trademark of MICROCOMPATIBLES, Inc., Silver Spring, MD) is a popular standalone graphics library that drives a microcomputer's screen and pen plotters. OMNIPLOT (S) produces screen graphics to preview plots; a hard copy can be obtained via a dump of the screen to a dot matrix printer. OMNIPLOT (P) produces a similar set of graphics commands for pen plotters with high resolution.

The following graphic formats are supported: tabular, bar, pie, contour, and 3-D wire frame. Choices include: pen speed, paper size, text size, plotting region, marker symbol type, line type, cubic spline interpolation to data, least squares fit to data, standard plots, semilog plots, log-log plots, gridding, tick mark label format, axis labels, plot labels, number and numerical value of contours, mesh size defining contour data, contour labels, and choice of size and viewing direction for 3-D plot. Data are entered in response to menu prompts or imported from a diskette file prepared with a text editor such as IBM's EDLIN editor on the PC-DOS diskette or a BASIC program included with the package.

OMNIPLOT requires contour map and 3-D plot data to be defined on a regular rectangular grid. Z values on an NX by NY (column by row) mesh of data points are user-defined. A new line is started each time the entry for a new row begins. Individual Z values are separated by a space or comma. An interpolation program is provided for filling out a nonregular array of data points. Programs WELFLO and CONMIG contain an option to create a file of Z values with a format matching that required by OMNIPLOT which may be integrated with other required pieces of formatted data (characters, integers, and real numbers) utilizing the DOS EDLIN Transfer (T) command.

OMNIPLOT supports the IBM PC and fully IBM-compatible PCs. Screen dumps are to an IBM dot matrix or compatible printer. Hewlett-Packard (HP-GL) or Houston Instrument (DMP) X-Y pen plotters are supported. Codes can utilize 8087/80287 math coprocessors.

Miller (1981) and Poole et al. (1981) provide extensive libraries of BASIC program listings which are useful in writing analytical groundwater flow, contaminant migration, and graphics codes. The following scientific subjects are covered in those libraries: linear interpolation; curvilinear interpolation; linear regression; Nth Order regression; sorting; Bessel, Gamma, and Gaussian Error functions; simultaneous solution of linear equations – Cramer's Rule, Gauss elimination method, Gauss-Jordan elimination method, and Gauss-Seidel method; solution of equations by Newton's method; numerical integration – Trapezoidal Rule, Simpson's Rule, and Romberg method; and nonlinear curve-fitting.

2

Well Function Program Operation

In WELFUN (see Appendix A), the well function of interest, $W(u)$, $W(u,r/B)$, $W(u_A,u_B,Beta)$, $W_{fp}(u_f,r/m_b,b,c)$, or $W(u,r(P_V/P_H)^{1/2}/m,L/m,d/m,Lo/m,do/m)$, is user-defined by entering an appropriate integer number in response to a well function option prompt. The user may choose to combine any of the first four well functions mentioned above with $W(u,r(P_V/P_H)^{1/2}/m,L/m,d/m,Lo/m,do/m)$ by entering an appropriate number in response to another well function option prompt.

Well function variable values are user-defined in response to data base entry prompts. Well function values are then calculated using polynomial and other approximations, except in the case of $W(u_A,u_B,Beta)$ where values are interpolated from tables stored in the computer memory. Well function curves are divided into two or more convenient segments in polynomial approximation algorithms. Computation results are displayed automatically on the screen and upon the user's request on paper with the printer. The user may choose to calculate several values of any one well function, or values of several well functions, without rerunning the program by entering an appropriate number in response to an option prompt.

3

Well Functions

Analytical model equations simulating transient ground-water flow toward production or flowing wells and/or flow from injection wells contain complex well functions (see Hantush, 1964, pp. 309–325). There are several methods for generating values of these functions with various levels of calculation precision including: numerical integral evaluation, approximate inversion of Laplace transforms, polynomial approximations, and linear or curvilinear interpolation of tabled function values. A list of common well functions with function symbols and brief well and associated aquifer condition descriptions is presented in Table 1. BASIC program WELFUN calculates values of the most commonly used well functions $W(u)$, $W(u,r/B)$, $W(u_A,u_B,Beta)$, $W_{fp}(u_f,r/m_b,b,c)$, and $W(u,r(P_V/P_H)^{1/2}/m,L/m,d/m,Lo/m,do/m)$ based on polynomial and other approximations and linear interpolation of tabled values.

Reed (1980, pp. 57–106) presents FORTRAN programs for generating precise values of the following well functions listed in Table 1: $W(u)$, $W(u,r(P_V/P_H)^{1/2}/m,L/m,d/m,Lo/m,do/m)$, $W(u,r/B)$, $W(u,Gamma)$, $W(Lambda,r_w/B)$, and $W(u_i,Rho_i)$. These numerically intensive programs evaluate integrals by using Gaussian-Laguerre quadrature formulas, series expansions, Simpson's Rule, Trapezoidal Rule, recurrence relation and polynomial approximation techniques, rational approximations, and asymptotic series (see IBM System/360 Scientific Subroutine Package; Watson, 1958; Krylov, 1962; Miller, 1957; and Shad/Chen/Frank,

Table 1.   Common Well Functions

Symbol	Well and Aquifer Description
$W(u)$	Production well, nonleaky artesian
$W(Lambda)$	Flowing well, nonleaky artesian
$W(u,S,Rho)$	Production well with wellbore storage, nonleaky artesian
$W(u_i,Rho_i)$	Slug injection well with wellbore storage, nonleaky artesian
$W(u_d,Rho_d)$	Pressurized injection well, nonleaky artesian
$W[u,r(P_V/P_H)^{1/2}$ /m,L/m,d/m,Lo/m, do/m]	Partially penetrating production well impact, nonleaky artesian
$W_{fp}(u_f,r/m_b,b,c)$	Production well, fractured rock
$W(u,r/B)$	Production well, leaky artesian
$W(Lambda, r_w/B)$	Flowing well, leaky artesian
$W(u,Gamma)$	Production well, leaky artesian with aquitard storativity
$W(u_A,u_B,Beta)$	Production well, water table with slow gravity yield

1964). FORTRAN subroutines for calculating values of the following Bessel functions are also presented by Reed (1980): $K_0(X)$, $K_1(X)$, $J_0(X)$, $J_1(X)$, $Y_0(X)$, and $Y_1(X)$.

A listing of a FORTRAN program for calculating precise values of the well function $W(u_A,u_B,Beta)$ is available (Neuman, 1975b). This program evaluates a complex integral numerically using a self-adapting Simpson's Rule algorithm. Cobb, et al. (1982) presents a FORTRAN numerical program for generating precise values of the well function $W(u,r/B)$.

Moench and Ogata (1984, pp. 146–170) developed Laplace transform solutions for the following complex well and aquifer conditions beyond those covered in Table 1: two interconnected aquifers with aquitard storativity, aquifer overlain by water table aquitard, partially penetrating well in leaky artesian aquifer with aquitard storativity, and large diameter well with wellbore storage in leaky artesian aquifer with aquitard storativity. The numerical inversion algorithm described by Stehfest (1970) can be used to generate precise values of the complex well functions associated with these solutions (see Moench and Ogata, 1984, pp. 150-151). A FORTRAN listing of a typical application of the Stehfest algorithm is given by Da Prat (1981). Dougherty and Babu (1984, pp. 1116-1122) present a numerical technique for generating precise slug test analysis well function values with fully or partially penetrating wells, wellbore storage, skin, and a double-porosity fractured rock aquifer. This technique also involves the Stehfest algorithm.

Sauveplane (1984, pp. 197-215) developed analytical equations for generating approximate values of the well functions $W(u)$ and $W_{fp}(u_f, r/m_b, b, c)$ listed in Table 1 and an approximate equation for estimating values of the well function associated with two interconnected aquifers and aquitard storativity. These solutions of well problems use Schapery's (1961) technique of approximate inversion of derived functions in the Laplace plane. Approximate analytical expressions for the well function $W(u, r/B)$ listed in Table 1 are provided by Hantush and Jacob (1955, pp. 95-100) and Hunt (1983, pp. 136-140).

Clark (1987) provides listings of BASIC programs for generating approximate values of well functions $W(u)$ and $W(u, r/B)$, first and second kind of zero order Bessel functions, the factorial function, and the Gaussian error function. Kinzelbach (1986, pp. 225-226) presents a BASIC program for generating values of $W(u, r/B)$ by performing required integration numerically with Simpson's Rule. Newton's successive approximation method (see Miller, 1981, pp. 203-223) is used in a BASIC program listed by Clark (1987, pp. 3.1-3.12), which calculates u when $W(u)$ is known.

Graphical microcomputer-assisted well function curve matching techniques are available for pumping test data analysis (Dansby, 1987, pp. 1523-1534). Drawdown or recovery values are plotted on the microcomputer screen and an appropriate well function curve is selected, overlain, and matched to the pumping test data directly on the screen. The hydraulic properties of the aquifer system are automatically estimated with curve match data. Well functions W(u), W(u,S,Rho), W(u$_i$,Rho$_i$), W(u,r/B), and W(u$_A$,u$_B$,Beta) are supported in software called "Graphical Well Analysis Package" distributed by Groundwater Graphics, San Diego, through the National Water Well Association (NWWA), as listed in Ground Water Report — A Catalog for the Industry and Others 1987/1988, p. 48. IGWMC maintains a list of other available software for aquifer test analysis. BASIC programs for pumping test design and analysis are provided by Walton (1987, pp. 105-162).

Polynomial and Other Approximations

Polynomial approximations of well functions W(u) and W(u,r/B) listed in Table 1 with precisions acceptable for most field applications are as follows (see Abramowitz and Stegun, 1964; Sandberg et al.,1981; Wilson and Miller, 1978):

W(u)
>   when $0 < u \leq 1$

$$W(u) = -\ln u + a_0 + a_1 u + a_2 u^2 + a_3 u^3 + a_4 u^4 + a_5 u^5$$

(1)

where

$$a_0 = -0.57721566 \qquad a_3 = 0.05519968$$
$$a_1 = 0.99999193 \qquad a_4 = -0.00976004$$
$$a_2 = -0.24991055 \qquad a_5 = 0.00107857$$

>   when $1 < u < \infty$

$$W(u) = [(u^4 + a_1u^3 + a_2u^2 + a_3u + a_4)/(u^4 + b_1u^3 + b_2u^2 + b_3u + b_4)]/uexp(u) \qquad (2)$$

where

$$
\begin{aligned}
a_1 &= 8.5733287401 & b_1 &= 9.5733223454 \\
a_2 &= 18.0590169730 & b_2 &= 25.6329561486 \\
a_3 &= 8.6347608925 & b_3 &= 21.0996530827 \\
a_4 &= 0.2677737343 & b_4 &= 3.9584969228
\end{aligned}
$$

According to Huisman and Olsthoorn (1983, p. 68), for $u < 0.25$ $W(u) = \ln(0.78/u)$ has an error of less than 1 percent. If $u > 1$ $W(u) = \exp(-1.2u - 0.60)$. If $u > 0.02$ $W(u,r/r_w) = W(u)$.

$W(u,r/B)$ (T.A. Prickett, oral communication, 1981)

when $r/B = 0$

$$W(u,r/B) = W(u) \qquad (3)$$

when $r/B > 0$ and $(r/B)^2/4u > 5$

$$W(u,r/B) = 2K_0(r/B) \qquad (4)$$

when $r/B > 0$ and $(r/B)^2/4u < 5$ and $u < 0.05$ and $u > 0.01$ and $r/B < 0.1$

$$W(u,r/B) = W(u) - [(r/B)/(4.7u^{0.6})]^2 \qquad (5)$$

when $r/B > 0$ and $(r/B)^2/4u \leq 5$ and $u \leq 0.9$ and $u \geq 0.05$ and $u > (r/B)/2$

$$W(u,r/B) = W(u) - [(r/B)/(4.7u^{0.6})]^2 \qquad (6)$$

when $r/B > 0$ and $(r/B)^2/4u \leq 5$ and $u \leq 0.01$ and $r/B \geq 0.1$

$$W(u,r/B) = 2K_0(r/B) - W[(r/B)^2/4u]I_0(r/B) \qquad (7)$$

when $r/B > 0$ and $(r/B)^2/4u \leq 5$ and $u \geq 0.05$ and $u \leq (r/B)/2$

$$W(u,r/B) = 2K_0(r/B) - 4.8 \times 10^E \qquad (8)$$

where

$$E = -(1.75u)^{-0.448(r/B)}$$

when $r/B > 0$ and $(r/B)^2/4u \leq 5$ and $u > 0.9$

$$W(u,r/B) = 1.5637\exp(-a-c/a)/a + 4.54\exp(-b-c/b)/b \quad (9)$$

where $a = u + 0.5858$ $b = u + 3.414$ $c = (r/B)^2/4$

when $r/B > 2$

$$W(u,r/B) = (\pi B/2r)^{0.5}\exp(-r/B)\text{erfc}\{-[(r/B) - 2u]/(2u^{0.5})\}$$
$$(10)$$

Other equations for $W(u,r/B)$ are presented by Case and Addiego (1977, pp. 393–397), Hunt (1977, pp. 179–183), and Streltsova (1988, p. 86).

Polynomial approximations for Bessel functions $K_0(X)$ and $I_0(X)$ and the error functions $\text{erfc}(X)$ and $\text{erf}(X)$ contained in well function $W(u,r/B)$ are (see Abramowitz and Stegun, 1964):

$K_0(X)$

when $0 < X \leq 2$

$$K_0(X) = -\ln(X/2)I_0(X) - 0.57721566 + 0.42278420(X/2)^2$$
$$+ 0.23069756(X/2)^4 + 0.03488590(X/2)^6$$
$$+ 0.0026298(X/2)^8 + 0.00010750(X/2)^{10}$$
$$+ 0.00000740(X/2)^{12} \qquad (11)$$

when $2 < X < \infty$

$$K_0(X) = [1.25331414 - 0.07832358(2/X) + 0.02189568(2/X)^2$$
$$-0.01062446(2/X)^3 + 0.00587872(2/X)^4 - 0.00251540(2/X)^5$$
$$+ 0.00053208(2/X)^6]/X^{1/2}\exp(X) \qquad (12)$$

$I_0(X)$

when $-3.75 \le X \le 3.75$

$$I_0(X) = 1 + 3.5156229(X/3.75)^2 + 3.0899424(X/3.75)^4$$
$$+ 1.2067492(X/3.75)^6 + 0.2659732(X/3.75)^8$$
$$+ 0.0360768(X/3.75)^{10} + 0.0045813(X/3.75)^{12} \qquad (13)$$

erfc(X)

$$\text{erfc}(X) = 1/(1 + a_1X + a_2X^2 + \ldots a_6X^6)^{16} \qquad (14)$$

$$\text{erfc}(-X) = 1 + \text{erf}(X) \qquad (15)$$

where

$$\text{erf}(X) = 1 - 1/(1 + a_1X + a_2X^2 + \ldots a_6X^6)^{16} \qquad (16)$$

$$\text{erf}(-X) = -\text{erf}(X) \qquad (17)$$

$a_1 = 0.0705230784 \qquad a_4 = 0.0001520143$
$a_2 = 0.0422820123 \qquad a_5 = 0.0002765672$
$a_3 = 0.0092705272 \qquad a_5 = 0.0000430638$

An approximation of the well function $W_{fp}(u_f, r/m_b, b, c)$ is as follows (Sauveplane, 1984, pp. 201–204):

$$W_{fp}(u_f, r/m_b, b, c) = 2K_0(A^{0.5}) \qquad (18)$$

where

$$A = 2u_f + rc(2u_f)^{0.5}/(m_b b^{0.5})\tanh(D) \qquad (19)$$

$$D = m_b(2u_f)^{0.5}/(rb^{0.5}) \tag{20}$$

$$\tanh(D) = \sinh(D)/\cosh(D)$$

$$\sinh(D) = [\exp(D) - \exp(-D)]/2$$

$$\cosh(D) = [\exp(D) + \exp(-D)]/2$$

Hantush (1961, p. 90) developed the following equation for calculating values of the well function $W(u,r(P_V/P_H)^{1/2}/m,L/m,d/m,Lo/m,do/m)$:

$$W[u,r(P_V/P_H)^{1/2}/m,L/m,d/m,Lo/m,do/m] = 2m^2/[\pi^2(1-d)$$

$$(Lo - do)] \sum_{n=1}^{50} 1/n^2[\sin(n\pi L/m) - \sin(n\pi d/m)]$$

$$[\sin(n\pi Lo/m) - \sin(n\pi do/m)]$$

$$W[u,n\pi r(P_V/P_H)^{1/2}/m] \tag{21}$$

A technique for deriving well functions with wellbore storage based on well functions without wellbore storage, the principle of superposition, and the method of successive approximations is presented by Streltsova (1988, pp. 195-201). By trial and error, values of aquifer discharge into a well and drawdown in a well for selected values of variables are calculated taking into consideration discharge from storage within the wellbore and drawdown without wellbore storage. Well functions with wellbore storage are then calculated based on values of drawdown and aquifer discharge into the well and appropriate groundwater model equations.

WELFUN subprograms for calculating values of well functions $W(u)$, $W(u,r/B)$, $W(u,r(P_V/P_H)^{1/2}/m,L/m,d/m,Lo/m,do/m)$, and $W_{fp}(u_f,r/m_b,b,c)$ are based on Equations 1-21.

Interpolation of Tabled Values

Tabled values of common well functions are given in the following references: Reed (1980, pp. 7-53); Ferris et al.

(1962, pp. 96–97); Hantush (1964, pp. 313,322–324); and Neuman (1975a, pp. 323–333). If the variable values of interest are not tabled, then desired values of well functions can be interpolated based on available values. Interpolation in one dimension is used to interpolate a single variable well function such as W(u), and interpolation in two dimensions is used to interpolate a multivariable well function such as W(u,r/B). The precision of interpolation depends on the spacing of tabled values and the method of interpolation. Linear interpolation is less precise than curvilinear interpolation (see Press et al., 1986).

Lagrangian curvilinear interpolation is based on the following equation (see Lancaster and Salkauskas, 1986, pp. 36–40):

$$Y_0 = \{[(X_0 - X_2)(X_0 - X_3) \ldots (X_0 - X_n)]/$$

$$[(X_1 - X_2)(X_1 - X_3) \ldots (X_1 - X_n)]\} Y_1$$

$$+ \{[(X_0 - X_1)(X_0 - X_3) \ldots (X_0 - X_n)]/$$

$$[(X_2 - X_1)(X_2 - X_3) \ldots (X_2 - X_n)]\} Y_2 + \ldots$$

$$+ \{[(X_0 - X_1)(X_0 - X_2) \ldots (X_0 - X_{n-1})]/$$

$$[(X_n - X_1)(X_n - X_2) \ldots (X_n - X_{n-1})]\} Y_n \qquad (22)$$

where

$X_1, X_2, \ldots, X_n$ and $Y_1, Y_2, \ldots, Y_n$ are the X,Y coordinates of known points, X_0 is the X coordinate of the interpolated point, and Y_0 is the calculated value of the Y coordinate of the interpolated point.

A BASIC program for Lagrangian interpolation between known points on a user-defined curve is listed by Poole et al. (1981, pp. 84–85). BASIC programs for Lagrangian interpolation of user-defined well functions in one- and two-dimensions are provided by Clark (1987, pp. 5.1–5.14).

The precision of linear interpolation is acceptable for most field applications. Linear interpolation between two points X_1,Y_1 and X_2,Y_2 in a table of well function values involves the following equation (Wilkes, 1966):

$$Y_0 = Y_1 + C\Delta Y \qquad (23)$$

where

Y_0 = interpolated value
C = $(X_0 - X_1)/(X_2 - X_1)$
ΔY = $Y_2 - Y_1$
X_0 = particular value of X at which Y_0 is desired

A WELFUN subprogram that calculates values of the well function $W(u_A,u_B,Beta)$ for which there is no readily available approximation is based on Equation 23. In WELFUN, ordered well function tabled values are entered with DATA statements. In one-dimension interpolation, X_0 is user-defined and the tabled values are searched (Weinman and Kurshan, 1985, pp. 145–149) to identify the two points X_1,Y_1 and X_2,Y_2 adjacent to X_0. The value Y_0 is then calculated with Equation 23. Well functions with more than one variable require multiple data input, searches, and linear interpolations using these procedures.

Algebraic Summation

The partially penetrating impact well function is integrated with well functions for fully penetrating wells in WELFUN to obtain well functions under partially penetrating well conditions. According to the principle of superposition (see Bear, 1979, pp. 152–159), the well function for a partially penetrating well is the algebraic summation of the well function with fully penetrating conditions and the partially penetration impact well function. Likewise, the well function for an aquifer having a single boundary is the algebraic summation of the well function for fully or partially penetrating conditions and an image well function (see Stallman, 1963, p. 46).

4

Groundwater Flow Program Operation

In WELFLO (see Appendix B), the density of groundwater is assumed constant throughout the aquifer. Supported English units are: gpm, cfs, days, min, ft, sq ft, cu ft, gpd/ft, and gpd/sq ft. Unit requirements are cited in data base input statements. Some useful unit conversions are: 1 liter/s = 15.85 US gal/min, 1 m/s = 2.119x10^6 US gal/day/sq ft, 1 m^2/s = 6.954x10^6 US gal/day/ft, 1 cm/s = 2.121x10^4 US gal/day/sq ft, 1 ft/day = 7.481 US gal/day/sq ft, 1 m/day = 24.5 US gal/day/sq ft, 1 US gal/day/sq ft = 0.0408 m/day, 1 US gal/day/sq ft = 0.134 ft/day, 1 US gal/day/sq ft = 4.72x10^{-5} cm/sec, and 1 ft = 0.3048 m.

The user specifies aquifer conditions to be simulated by entering an appropriate number in response to an aquifer option prompt. The following conditions are supported: nonleaky artesian, leaky artesian, water table and nonleaky artesian fractured rock. It is assumed that aquitard storativity and delayed gravity yield are negligible. The user may specify fully or partially penetrating wells with or without wellbore storage by entering an appropriate number in response to a well option prompt. It is assumed that well discharge rates are constant if wells partially penetrate the aquifer and/or wellbore storage is simulated. The user specifies whether partial penetration impacts are to be calculated for the production well or an observation well by entering an appropriate number in response to another well option prompt.

The general data base is user-defined at the keyboard and displayed upon the user's request on paper by the printer. In response to a series of prompts, the number of simulation periods is specified by the user (must be < 26). Graphs of well discharge (+) and well recharge (–) versus time are drawn by the user for each production and injection well and combined into simulation period pump operation schedules. Based on these graphs, the duration of each simulation period is user-defined as the longest pump operation period during each simulation period. Simulation period durations are entered in ascending order. The differences between simulation period times must be small for high precision in wellbore storage impact calculations.

Any aquifer boundaries and discontinuities, real production and/or injection wells, and any boundary or discontinuity image wells are drawn to scale on a map. An area of interest which lies entirely within aquifer boundaries and/ or discontinuities is selected for drawdown calculation and display. A uniform grid is superposed over this area. Grid lines are indexed using the I (column), J (row) notation colinear with the X and Y directions, respectively. I coordinates increase left to right and J coordinates increase top to bottom. The origin of the grid is beyond the grid at the upper-left corner of the map. The X and Y coordinates of the upper-left grid node are user-defined in a manner so that confusing drawdowns or recoveries beyond boundaries and/or discontinuities are not displayed. The number of columns must be < 31 and the number of rows must be < 31. The number of columns or rows must be 10, 20, or 30 and the grid must be square if calculation results are to be used with GWGRAF. The X coordinate of the upper-left node and Y coordinate of the upper-left node are user-defined.

The number of active production, injection, and image wells during each simulation period (must be < 51), the X coordinate, Y coordinate, discharge rate, and radius of each well are user-defined. The number of observation wells located at grid nodes for which time-drawdown or recovery

tables are desired (must be < 26) and the I coordinates and
J coordinates of observation wells are user-defined.
Depending upon aquifer conditions, appropriate values of
some or most of the following aquifer system hydraulic
properties are user-defined: aquifer transmissivity, aquifer
storativity, aquifer specific yield, aquitard thickness,
aquitard vertical hydraulic conductivity, fissure horizontal
hydraulic conductivity, block vertical hydraulic conductivity, fractured rock aquifer thickness, storativity of fissured
portion of fractured rock aquifer, storativity of block portion of fractured rock aquifer, and half thickness of average
block unit. The user may choose to revise and print on
paper the general data base.

The coordinates of grid nodes are determined with data
on the known coordinates of the upper-left node. The distances between wells and grid nodes are calculated using
the Pythagorean equation. Well function variables are calculated using the general data base, and associated well
function values are determined using polynomial and other
approximations. Individual well drawdown and/or recovery
impacts at nodes are calculated using appropriate aquifer
model equations and combined for the time increment in
question. Drawdown or recovery inside a grid block may be
calculated based on the impacts at the four surrounding
block corner grid nodes with the Lagrange-interpolation
equation (Kinzelbach, 1986, p.68). Nodal drawdown or
recovery tables of computation results for the selected area
of interest are displayed automatically on the screen and
upon the user's request on paper by the printer. The user
may choose to create a sequential data file of computation
results for export to graphics programs GWGRAF,
SURFER, and OMNIPLOT.

The well penetration data base is entered from the keyboard in response to input prompts if the user earlier chose
to simulate partially penetrating wells. Drawdown with full
penetration in the well in question, production well discharge rate, aquifer thickness, and well geometry are user-
defined. The well partial penetration impact is then calcu-

lated using the partial penetration well function approximation. Drawdown with partial penetration impacts is displayed automatically on the screen and upon the user's request on paper by the printer. The user may choose to repeat the partial penetration subprogram to calculate partial penetration impacts in additional wells without rerunning the program.

The wellbore storage data base is entered from the keyboard in response to input prompts if the user chose to simulate wellbore storage. Constant production well discharge during the present simulation period, drawdown with or without partial penetration and without wellbore storage at the end of present simulation period, and constant well and pump geometry are user-defined. If the well in question is a production well, drawdown without wellbore storage and with a finite diameter instead of an infinitesimal diameter is calculated using tabled values of the finite diameter well function. Discharge from the aquifer with wellbore storage is then calculated using the method of successive approximations. Drawdown without wellbore storage is adjusted for wellbore storage impacts using the ratio discharge from the aquifer versus total discharge from the production well. The results of computations are displayed automatically on the screen and upon the user's request on paper by the printer.

Tables of observation well time-drawdown or recovery data are displayed automatically on the screen and upon the user's request on paper by the printer after drawdowns or recoveries for all time increments have been calculated.

5

Groundwater Flow Simulation

Although not as versatile as numerical models, analytical models are useful in simulating groundwater flow in aquifer systems with low to moderate complexities. Recognized departures from ideal homogeneous and isotropic conditions and straight-line boundary demarcations assumed in analytical models do not necessarily dictate that they be rarely used. With appropriate recognition of complicated field situations, there are many practical ways of circumventing analytical model limitations using equivalent hydraulic property or aquifer cross section, incremental, and successive approximation techniques (see Walton, 1984a, pp. 301–302).

Assuming sound professional judgment, transient groundwater flow under various well and aquifer conditions can be simulated with analytical models consisting of well functions and their associated well hydraulics equations, the method of images (see Bear, 1979, pp. 356–366; Streltsova, 1988, pp. 210–253), and the principle of superposition (see Bear, 1979, pp. 152–159). Variable production well discharge rates may be simulated by placing several production wells on top of one another at a grid node or by using the numerical pumping rate change program listed by Clark (1987, pp. 7.1–7.16). Drawdown equations with linear, parabolic, polynomial, exponential, and periodical variable discharge rates are presented by Streltsova (1988, pp. 101–152).

Such features as mines, drains, pits, ponds, mounds, and

streams may be simulated by arrangement of closely spaced wells in lines, rectangles, circles, and irregular patterns (see Figure 1). Flow to mines and drains with fixed drawdown may be estimated by using the method of successive approximations. Drawdown due to first trial well discharges simulating mine drainage is calculated and compared to required drawdown. If the comparison is favorable, the well discharges are declared valid; otherwise, well discharges are revised and a second trial drawdown is calculated and compared to this required drawdown. This process is repeated until the comparison is favorable.

Clark (1987, pp. 2.1–2.28, 4.1–4.49, and 7.1–7.16) lists BASIC programs for calculating drawdown in nonleaky artesian and leaky artesian infinite, one-boundary, and strip aquifers with multiple discharge rates. Walton (1984b) lists eight BASIC programs for one-, two-, and three-layered aquifer systems, a circular recharge area, and stream depletion. A collection of 18 articles from the journal *Ground Water* including microcomputer program listings pertaining to pumping tests, optimal well discharge, and well drawdown is distributed by the National Water Well Association (NWWA). Sandberg et al. (1981) provide analytical model equations for simulating source/sink flow rates with drawdown as the given, steady-state drawdown around finite line sinks, finite line sinks with nonsteady conditions, and steady state flow to line sources or sinks with drawdown as the given.

Boundaries and Discontinuities

Well hydraulics equations assume that the aquifer is infinite in areal extent. Finite aquifer conditions may be simulated by replacing hydrogeologic boundaries with imaginary wells which produce the same disturbing effects as the boundaries. The image well theory may be summarized as follows: the effect of a barrier boundary on the drawdown in a well, as a result of pumping from another well, is the

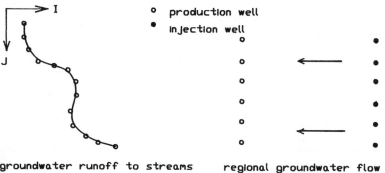

production well

injection well

groundwater runoff to streams

regional groundwater flow

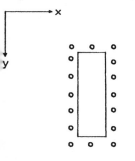

dewatering trench

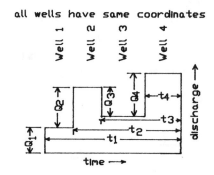

all wells have same coordinates

variable well discharge

pit, pond, or lagoon seepage

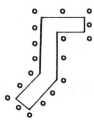

mine drift drainage

Figure 1.   Groundwater flow models

same as though the aquifer were infinite and a like discharging well were located across the real boundary on a perpendicular thereto and at the same distance from the boundary as the real pumping well. For a recharge boundary, the principle is the same except the image well is assumed to be recharging the aquifer instead of pumping from it.

Boundary problems are thereby simplified to consideration of an infinite aquifer in which real and image wells operate simultaneously at the same discharge or recharge rate. Successive reflections on boundaries are considered when aquifers are delimited by two or more boundaries (see Figure 2). In the case of an aquifer discontinuity (partial boundary), an image well is placed as usual but the image well strength is calculated with the following equation (Muskat, 1937; Streltsova, 1988, pp. 246-251; Fenske, 1984, pp. 125-145):

$$Q_i = Q_p D_t \tag{24}$$

where

$$D_t = (T_p - T_d)/(T_p + T_d) \tag{25}$$

Q_i = image well strength in gpm

Q_p = production well discharge rate in gpm

T_p = aquifer transmissivity at production well in gpd/ft

T_d = aquifer transmissivity beyond discontinuity in gpd/ft

Heterogeneities and Stratification

Aquifer anisotropic and heterogeneous and aquitard stratification impacts may be simulated by using average hydraulic properties instead of complex horizontal and vertical hydraulic properties variations in conceptual models. Average hydraulic conductivities may be calculated with

due to recharge boundary image wells. Stated as an equation

$$s_T = s + s_{wl} + s_p + s_d + s_b - s_r \qquad (28)$$

Values of s, s_p, s_b, and s_r are calculated in WELFLO utilizing four common analytical models and associated equations.

Model 1 (Theis, 1935, pp. 519–524) assumes that flow is entirely horizontal and radial and that wells fully penetrate the aquifer. There are no vertical components of flow and no wellbore storage. The uniformly porous aquifer is overlain and underlain by aquicludes with negligible vertical hydraulic conductivity. The nonleaky artesian aquifer is homogeneous, isotropic, infinite in areal extent, and constant in thickness throughout. Wells have infinitesimal diameters and the discharge rate is constant. There are no boundaries or discontinuities. Model 1 simulates water table conditions when delayed gravity yield impacts are negligible and drawdown is small relative to the initial saturated aquifer thickness. The Model 1 equation is:

$$s = 114.6QW(u)/T \qquad (29)$$

where

$$u = 1.87r^2S/(Tt) \qquad (30)$$

Q = discharge rate in gpm

$W(u)$ = well function, dimensionless

T = aquifer transmissivity in gpd/ft

r = distance from production well in ft

S = aquifer storativity as a decimal

t = time after pumping started in days

Rushton and Redshaw (1979, pp. 237–266) list a short numerical FORTRAN program which simulates water table conditions with slow gravity yield and decreased transmissivity with water table decline. A BASIC version of that program is listed by Walton (1987, pp.106–112).

Model 2 (Hantush and Jacob, 1955, pp. 95–100) assumes wells fully penetrate a leaky artesian aquifer overlain by an aquitard and underlain by an aquiclude. Overlying the aquitard are deposits (source bed) in which there is a water table. The aquifer is homogeneous, isotropic, infinite in areal extent, and constant in thickness throughout. Flow lines are assumed to be refracted a full right angle as they cross the aquitard-aquifer interface. The aquitard is assumed to be more or less incompressible so that water released from storage therein is negligible. Drawdown in the source bed is negligible, production and observation wells have infinitesimal diameters and no wellbore storage, and the discharge rate is constant. There are no boundaries or discontinuities. Neuman and Witherspoon (1969, pp.810, 821) indicate that the use of Model 2 is justified when $r/(4m)[P_c S_c m/(P_H S m_c)]^{1/2} < 0.1$ and $t < 1.08 \times 10^3 m_c S_c/P_c$. Under these conditions, change in storage in the source bed are negligible. The Model 2 equation is:

$$s = 114.6QW(u,r/B)/T \qquad (31)$$

where

$$B = (Tm_c/P_c)^{1/2} \qquad (32)$$

m_c = aquitard thickness in ft

P_c = aquitard vertical hydraulic conductivity in gpd/sq ft

$W(u,r/B)$ is a dimensionless well function

Model 3 (Boulton and Streltsova, 1977, pp. 257–270) assumes a fractured rock aquifer overlain and underlain by aquicludes. The nonleaky artesian aquifer consists of porous blocks of irregular size and shape, and adjoining and interconnected fissures. The aquifer is simulated by a set of regular porous block-and-fissure units made up of a horizontal porous block and a planar fissure. The thickness of the fissure is small compared with that of the porous block. The nonleaky artesian aquifer is heterogeneous, infinite in areal extent, and constant in thickness throughout. Wells have infinitesimal diameters and no wellbore storage. The production well is cased through porous blocks. The Model 3 equation is (see Streltsova, 1988, pp. 313–319, 381–391):

$$s_f = 114.6 Q W_{fp}(u_f, r/m_b, b, c)/T_f \qquad (33)$$

where

$$u_f = 1.87 r^2 S_f/(T_f t) \qquad (34)$$

$$b = P_b S_f/(S_b P_f) \qquad (35)$$

$$c = P_b/P_f \qquad (36)$$

$$T_f = P_f m \qquad (37)$$

s_f = drawdown in fissures in ft

S_f = fissure storativity as a decimal

T_f = fissure horizontal transmissivity in gpd/ft

T_b = block vertical transmissivity in gpd/ft

S_b = block storativity as a decimal

m_b = half dimension (half thickness) of average block unit in ft

m = fractured rock aquifer thickness in ft

P_b = block vert. hydr. conductivity in gpd/sq ft

P_f = fissure horiz. hydr. conductivity in gpd/sq ft

$W_{fp}(u_f, r/m_b, b, c)$ is a dimensionless well function

Moench (1984, pp. 831–846) presents an analytical solution in the Laplace plane evaluated by numerical inversion for flow to a well of finite diameter with wellbore storage in a fractured rock aquifer. Wellbore storage impacts are shown to be dominant and to overshadow double-porosity impacts during early time periods.

Values of s_p are calculated with the Model 4 equation (Hantush, 1961, pp. 83–98), assuming the well discharge rate is constant during the simulation period:

$$s_p = 114.6QW[u, r(P_V/P_H)^{1/2}/m, L/m, d/m, Lo/m, do/m]/T \quad (38)$$

where

P_V = aquifer vertical hydraulic conductivity in gpd/sq ft

P_H = aquifer horiz. hydraulic conductivity in gpd/sq ft

m = aquifer thickness in ft

L = distance from aquifer top to prod. well base in ft

d = distance from aquifer top to prod. well screen top in ft

Lo = distance from aquifer top to obs. well base in ft

do = distance from aquifer top to obs. well screen top in ft

The effects of a damaged or enhanced zone surrounding a production well (skin; see Streltsova, 1988, pp. 75–78) may be simulated through the proper selection of the effective radius of the well. Well loss may be calculated with the following equation (Jacob, 1946, pp. 1047–1070):

$$s_{wl} = CQ^2 \tag{39}$$

where

s_{wl} = drawdown component due to well loss in ft

C = well loss coefficient in sec^2/ft^5

Q = production well discharge rate in cfs

Values of s_d may be calculated with the following equation (Jacob, 1946):

$$s_a = s_o - s_o^2/(2m) \tag{40}$$

where

s_a = drawdown with dewatering effects in ft

s_o = drawdown without dewatering effects in ft

m = initial aquifer thickness in ft

Wellbore Storage

Whether or not wellbore storage should be simulated may be determined with the following equation (see Driscoll, 1986, p. 566):

$$t_s = 5.4 \times 10^5 (r_w^2 - r_c^2)/T \tag{41}$$

where

r_w = production well effective radius in ft

r_c = pump-column pipe radius in ft

T = aquifer transmissivity in gpd/ft

t_s = time after pumping started beyond which wellbore storage impacts are negligible (less than 1% of drawdown values) in minutes

Wellbore storage is approximated in WELFLO with an iterative procedure described by Huyakorn and Pinder (1983, pp. 127–128). The procedure assumes the well discharge rate and well geometry are constant during simulation periods and drawdown under water table conditions is negligible in comparison to the aquifer thickness. The procedure uses the principle of superposition. Well discharge is the sum of the discharge derived from the aquifer and the discharge derived from storage within the wellbore. Discharge from the aquifer increases exponentially with time (Streltsova, 1988, p. 51). However, in the approximation it is assumed that discharge from the aquifer varies linearly during short pumping periods. Drawdown with wellbore storage at the end of a time increment is the drawdown with wellbore storage at the beginning of the time increment, plus the drawdown without wellbore storage during the time increment, multiplied by the average aquifer discharge, divided by the well discharge.

The equation relating the constant discharge from the production well, average discharge derived from the aquifer, and discharge derived from storage within the wellbore is as follows (see Huyakorn and Pinder, pp. 127–128):

$$Q = Q_A + \pi(r_w{}^2 - r_c{}^2)s_{ws}7.48/\Delta t \qquad (42)$$

where

Q = constant discharge from production well during simulation period in gpm

Q_A = average discharge derived from aquifer during simulation period (sum of aquifer discharges at simulation period start and end divided by 2) in gpm

r_w = production well effective radius in ft

r_c = pump column-pipe radius in ft

s_{ws} = drawdown with wellbore storage impacts during simulation period in ft

Δt = time at end of present simulation period in minutes

The successive bisection iterative procedure used in WELFLO (see Clark, 1987, pp. 3.8–3.12) starts with a first trial value of s_{ws} based on the analytically calculated drawdown without wellbore storage impacts. If needed, early drawdown in the production well due to pumping the well itself is calculated with the finite diameter well function (see Streltsova, 1988, pp. 45–49). Equation 42 is used to calculate a first trial value of Q_A. A second trial value of s_{ws} is calculated by multiplying drawdown without wellbore storage impacts by the ratio first trial value of Q_A, divided by the discharge from the well. The first and second trial values of s_{ws} are compared and if the difference exceeds an error tolerance the iteration is repeated. If the difference is less than or equal to an error tolerance, the estimated values of Q_A and s_{ws} are declared valid.

Complex wellbore storage functions may be calculated with values of s_{ws}. For example, suppose it is desirable to determine values of the leaky artesian wellbore storage function. Values of drawdown (s) without wellbore storage are calculated based on leaky artesian Model 2 Equations 31 and 32 and selected values of u and r/B. Corresponding values of drawdown with wellbore storage (s_{ws}) are then calculated using the iterative procedure and Equation 42. Corresponding values of the leaky artesian wellbore storage function $W(u,r/B,S,r/r_w)$ are finally calculated with the equation $W(u,r/B,S,r/r_w) = s_{ws}T/(114.6Q)$.

6

Contaminant Migration
Program Operation

In CONMIG (see Appendix C), the user specifies the number of simulation periods (must be < 26) for which contaminant concentration distribution is to be calculated. Time schedules for contaminant point sources are prepared by the user. Simulation period durations are set equal to the longest time schedule in each period.

Milligram-liter-pound-day units are supported in CONMIG. Required units are specified in interactive data base input statements. Useful unit conversions are: 1 gram = 2.205×10^{-3} pounds, 1 cu meter = 2.642×10^{2} US gal, 1 cu ft = 7.48 US gal, 1 US gal = 8.3453 pounds, 1 liter = 0.264 US gal, 1 US gal = 3.785 liters, 1 pound/cu ft = 1.602×10^{4} mg/L, and 1 cu meter = 264.2 US gal.

Contaminant point sources and other features within the area of concern are drawn to scale on a map by the user. A uniform square grid is superposed over the map. Grid lines are indexed using the I (column), J (row) notation colinear with the X and Y directions, respectively. I coordinates increase left to right and J coordinates increase top to bottom. The origin of the grid is at or beyond the upper-left corner of the map. The number of grid columns (must be < 31), the number of grid rows (must be < 31), the grid spacing, and the coordinates of the upper-left grid node are user-defined. The number of columns or rows must be 10, 20, or 30 and the grid must be square if calculation results are to be used with GWGRAF.

The user specifies the aquifer actual porosity, effective porosity, thickness, longitudinal dispersivity, transverse dispersivity, and seepage velocity (Darcy velocity divided by effective porosity). The number of point sources (must be < 31) and whether each source is continuous or slug are user-defined. The user may choose to express the contaminant load of each point source in the case of a continuous source as mass injection rate in pounds per day, or as the injection rate in gpd with solute concentration in mg/L and in the case of a slug source as injected mass in pounds or as volume of injected mass in gal with solute concentration in mg/L. The coordinates of each point source are user-defined.

The user may choose to simulate adsorption and/or radioactive decay during all simulation periods and to specify the bulk density of the aquifer skeleton, aquifer distribution coefficient, and the half-life of the radionuclide. It is assumed that adsorption parameters are the same during all simulation periods and that radioactive decay starts at the beginning of the simulation periods. The constant retardation factor and decay factor are calculated. The number of monitor wells located at grid nodes for which time-concentration tables are desired (must be < 26) are user-defined. The coordinates of the monitor wells are specified and monitor well concentrations are initialized. A hard copy of the data base is printed upon the user's request on paper. The user may choose to revise the data base.

Grid node coordinates, variables in contaminant migration equations, continuous source functions, and concentrations at nodes are calculated with appropriate contaminant migration model equations. Nodal computation results are displayed automatically on the screen and upon the user's request on paper by the printer in tabular form. The user may choose to create a sequential data file of computation results for export to graphics programs GWGRAF, SURFER, or OMNIPLOT. Time-concentration computation results are displayed automatically on the screen and upon the user's request on paper by the printer in tabular form.

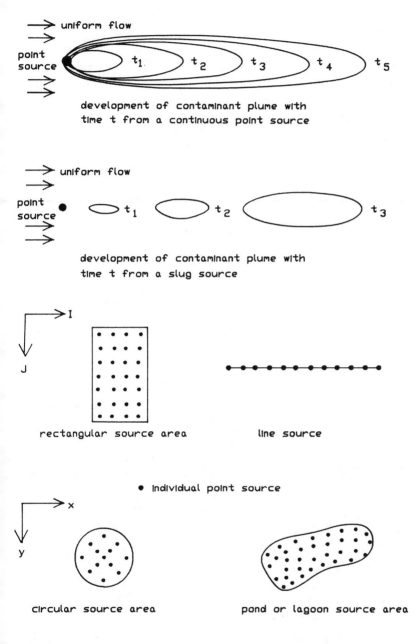

Figure 3. Contaminant migration models.

Point Source Models

The common two-dimensional equation governing contaminant migration in uniform one-directional flow from a continuous point source without adsorption or radioactive decay employed in program CONMIG is (Wilson and Miller, 1978, pp. 504–505):

$$C = 1.064 \times 10^{-2} C_0 Q_s \exp(x/B_d) W_s(u_d, r_d/B_d)/[mn(D_L D_T)^{1/2}]$$
(43a)

or

$$C = 1.275 \times 10^3 M_s \exp(x/B_d) W_s(u_d, r_d/B_d)/[mn(D_L D_T)^{1/2}]$$
(43b)

where

$$u_d = r_d^2/(4D_L t) \qquad (44)$$

$$r_d^2 = x^2 + y^2(D_L/D_T) \qquad (45)$$

$$B_d = 2D_L/v_s \qquad (46)$$

$$D_L = A_L v_s \qquad (47)$$

$$D_T = A_T v_s \qquad (48)$$

$W_s(u_d, r_d/B_d)$ has the same value as the function $W(u, r/B)$

C_0 = difference between solute concentration injected into aquifer and native aquifer solute concentration in mg/L

C = change in aquifer solute concentration due to solute injection in mg/L

Q_s = solute injection rate in gpd

x, y = cartesian coordinates of monitoring wells in ft

m = aquifer thickness in ft

n = aquifer effective porosity as a decimal

M_s = mass injection rate in pounds per day

t = time after injection started in days

v_s = seepage velocity without adsorption in ft/day

A_L = longitudinal dispersivity without adsorption in ft

A_T = transverse dispersivity without adsorption in ft

The common two-dimensional equation governing contaminant migration in uniform one-directional flow from a slug point source without adsorption or radioactive decay employed in program CONMIG is (Hunt, 1983, p. 134):

$$C = 1.064 \times 10^{-2} C_0 V_c \exp - \{[(x - v_s t)^2/(4 A_L v_s t)]$$
$$+ y^2/(4 A_T v_s t)\}/[mn v_s (A_L A_T)^{1/2} t] \qquad (49a)$$

or

$$C = 1.275 \times 10^3 M_c \exp - \{[(x - v_s t)^2/(4 A_L v_s t)]$$
$$+ y^2/(4 A_T v_s t)\}/[mn v_s (A_L A_T)^{1/2} t] \qquad (49b)$$

where

M_c = injected mass in pounds

V_c = volume of injected mass in gal

Adsorption and Radioactive Decay

Equations 43 and 49 and the retardation factor (see Marsily, 1986, pp. 256–257) are used in program CONMIG to simulate contaminant migration with adsorption. A_L, A_T, v_s, and C are divided by the retardation factor defined as:

$$R_d = 1 + [(D_{bs}/n)K_d] \qquad (50)$$

where

R_d = retardation factor as a decimal

D_{bs} = bulk density of dry aquifer skeleton in g/cm^3

n = aquifer effective porosity as a decimal

K_d = distribution coefficient in ml/g

Dividing the contaminant concentration by the retardation factor reflects the fact that part of the contaminant mass is adsorbed onto the aquifer matrix and does not contribute to dissolved concentration (see Kinzelbach, 1986, p. 204). In decontamination with desorption, contaminant mass is desorbed from the aquifer matrix and contributes to dissolved concentration.

Equation 50 assumes that single contaminant reactions are fast and reversible, the isotherm is linear, and there is no mixture of contaminants. The complexities, uncertainties, and limitations surrounding dispersivity and inherent in the simulation of sorption are discussed in detail by Anderson (1984), Cherry et al. (1984), and Miller and Weber (1984).

Radioactive decay is simulated in program CONMIG with the following equation (see Marsily, 1986, p. 265; Codell et al., 1981):

$$C_r = Ce^{-Zt} \tag{51}$$

where

$$Z = 0.693/h_1 \tag{52}$$

C_r = concentration of solute with radioactive in mg/L

C = concentration of solute without radioactive decay in mg/L

t = time after radioactive decay started in days

h_1 = half-life of substance in days

8

Graphics Program Operation

In GWGRAF (see Appendix D), the user may interpolate a square grid pattern of XYZ values from scattered surface data, create a time-drawdown semilog graph with best-fit line, create a time-drawdown or time-concentration arithmetic graph, or create a contour map of XYZ values. GWGRAF provides a convenient prompt system for user-entry of the data base from the keyboard or an "import" option to load the data base from a diskette file. To import a file, the user must first create an ASCII (text) file with WELFLO or CONMIG. The formatted file may also be created with an editor or word processor. GWGRAF's output may be directed to the screen, to a diskette file, or to a dot matrix printer, assuming that the DOS command GRAPHICS is entered by the user. The program requires a high resolution monitor (640 by 200) and an IBM Color Graphics Adapter or compatible device. A fatal error will occur with other resolution monitors and/or devices. Decisions concerning axes, tick marks, scales, labels, data point symbols, sizes, titles, and lines for trend graphs and contour maps are made internally in the program to simplify user-entry.

The following data are user-defined in response to input prompts in the subprogram to interpolate a square grid pattern of values from scattered surface data: number of known XYZ data points (must be > 5 and < 31), Z value for each data point, X coordinate for each data point, Y-coordinate for each data point, number of grid columns (must be $<$ 31), number of grid rows (must be $<$ 31), grid spacing, X coordinate of upper-left grid node, and Y coordinate of upper-left grid node (special requirements on pg. 48). The user may choose to revise the data base. Distances between grid nodes and data points are calculated using the Pythagorean equation and sorted. Weighting factors for the six closest data points to grid nodes are calculated, and Z values at grid nodes are interpolated using the weighting factors and the equation of a straight line.

Nodal computation results are displayed in tabular form automatically on the screen and upon the user's request on paper by the printer. The user may choose to create a sequential data file of computation results for export to graphic programs GWGRAF, OMNIPLOT, and SURFER.

In the subprogram to create a time-drawdown semilog graph, drawdown is plotted on the vertical axis and the log of time after pumping started is plotted on the horizontal axis. The maximum number on the horizontal axis is automatically set at 100000. The maximum number on the vertical axis is user-defined as a multiple of 10. The following data are user-defined in response to input prompts: production well discharge, distance between production and observation wells, number of known XY data points (must $<$ 31), time-coordinate of each data point, and drawdown-coordinate of each data point. It is assumed that the data points are for times when $u < 0.02$. A hard copy of the data base is made on paper upon the user's request by the printer. The user may choose to revise the data base.

Coordinates of the best-fit line through the data base are calculated by linear regression using the method of least squares. Values of aquifer transmissivity and storativity are calculated with best-fit line data and displayed automatically on the screen and upon the user's request on paper by the printer.

Graph axes, ticks, frame, data symbols, and best-fit line are displayed on the screen. A hard copy of the screen may be made by holding down the Shift key and pressing the

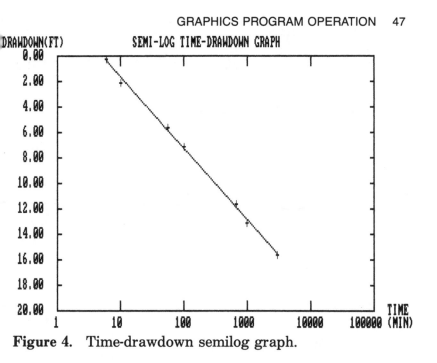

Figure 4.   Time-drawdown semilog graph.

PrtSc key (see Figure 4). After obtaining the hard copy, any key is pressed to end the subprogram.

The following data are user-defined to create a time-drawdown or time-concentration arithmetic graph: maximum number on the horizontal axis (must be ≥ 10 and < 10000), maximum number on the vertical axis (must be ≥ 10), number of known XY points (must be < 31), time coordinate of each data point, and drawdown or concentration coordinate of each data point. Drawdown or concentration are plotted on the vertical axis, and time is plotted on the horizontal axis. The user may choose to revise the data base.

The graph axes, ticks, frame, labels, data point symbols, and lines connecting data points are displayed automatically on the screen. The closer the data points, the smoother the curve through the data points will appear. A hard copy of the screen may be made by holding down the Shift key and pressing the PrtSc key (see Figure 5). After

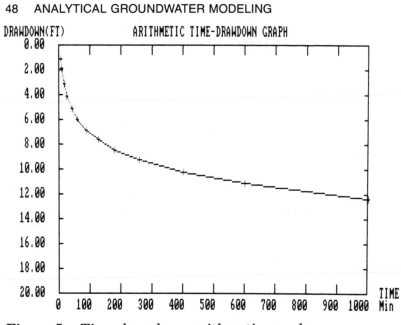

Figure 5.   Time-drawdown arithmetic graph.

obtaining the hard copy, any key is pressed to end the subprogram.

In the subprogram, to create a drawdown or concentration contour map, instructions are displayed automatically on the screen for subprogram operation. The user may choose to obtain a hard copy of these instructions. The user specifies a valid data base file name and the XYZ data base is imported from a diskette. The format of the data base must be as follows: $Z(1,1)$, $Z(2,1)$, . . . up to $Z(NC,1)$ where NC=number of columns, then, $Z(1,2)$, $Z(2,2)$, . . . up to $Z(NC,2)$. These entries continue through $Z(NC,NR)$ where NR=number of rows. It is assumed that the XYZ data base is in the form of a square grid.

The user specifies the number of grid columns (must be 10, 20, or 30), number of grid rows (must be 10, 20, or 30), number of contours to be displayed (must be < 11), grid spacing, and contour values (must be five digits or less). The number of grid columns and rows must match those in

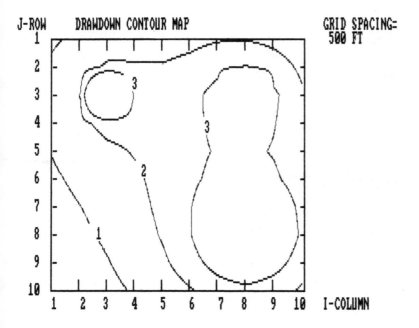

Figure 6.   Contour map graphics display.

the data base file. The user may choose to revise the data base.

The contour map axes, ticks, labels, and frame are displayed automatically on the screen. Closely spaced coordinates on contours are calculated using the straight line equation and linear interpolation between grid node Z values. Pixels (picture elements) at contour coordinates are turned on to display the contours. This process may take 10 minutes per contour if GWGRAF is not compiled because time consuming interpolations are performed through both I and J directions.

After the contours have been displayed, the user may label the contours interactively. A dot cursor is displayed on the screen. The cursor may be moved in small steps to the right, left, up, or down to the location of each contour label by pressing R, L, U, or D, respectively. The cursor algorithm turns on a pixel (dot cursor) with selected coordinates in the upper-left corner of the screen. If the cursor is not immediately visible, press R, L, U, or D until it is visible. The cursor coordinates are sequentially changed to represent increments of cursor movement, depending upon the cursor prompt letter chosen by the user. The pixel at the old cursor coordinates is turned off and the pixel at the new cursor coordinates is turned on, thereby simulating a moving cursor.

When the cursor is located at the lower-left corner of the contour label, the user presses the key N. The user is prompted to enter the contour value based on a cursory inspection of the data base in the lower-right portion of the screen outside the map area. The chosen number will briefly appear in the lower portion of the screen and then be automatically erased. Any contour line within the space occupied by a 5-digit number to the right of the lower-right corner of the contour label is erased. The number is then displayed at the contour label location. When the contour labeling is completed and a hard copy of the screen is obtained, the user presses the E key and then any other key to end the subprogram. The labeling algorithm is based on the cursor coordinates and the GET and PUT statements. A hard copy of the screen may be made by holding down the Shift key and pressing the PrtSc key (see Figure 6) provided the DOS command GRAPHICS has been previously entered by the user (see Appendix E and DOS manual).

9

Graphics Techniques

Output from programs WELFLO and CONMIG is in the form of numeric tables displayed on the screen and copied on paper by a printer, or stored in a file for export to a graphics program. Data listed in tables are usually more difficult to interpret than when printed in graphic form. GWGRAF is a medium quality screen/dot matrix printer graphics program that creates at the keyboard or imports a data base file and draws pictorial time-drawdown or time-concentration XY graphs and drawdown or concentration XYZ contour maps (see Figures 4–6). Screen graphics subprograms turn on appropriate patterns of pixels to produce pictures with points and lines (Hearn and Baker, 1983, pp. 31–53). Pressing simultaneously the Shift and PrtSc keys converts a screen's picture in the microcomputer's memory into a picture on a piece of paper (hard copy). Screens and dot matrix printers are "raster" devices which draw by producing or merging dots in a matrix (Sandler, 1984, pp. 55–89). Graphics programs often involve statistical equations and subprograms (see Poole et al. 1981).

The trend graph displays a single data series over a period of time or space interval. In trend graph programs (see Hearn and Baker, 1983, pp. 54–69; Korites, 1982, LIN/LIN, LOG/LIN, LOG/LOG, and LINFIT) axes are drawn together with scale interval ticks and labels. Data points are identified and lines or curves are drawn between points. Axes and the graph are provided with appropriate annotations. It is often desirable to draw a trend line or curve of best-fit that lies as close as possible to the largest number of points.

Contour maps represent paths of lines or curves connecting points having common values on a two-dimensional XY surface (see Jones, et al. 1986). Contouring requires interpolation between control points to define intermediate points; contour positions are located by interpolating between intermediate points. Contour lines are labeled, axes are drawn together with scale interval ticks and labels, and the map is provided with appropriate annotations. Microcomputer programs for creating contour maps are presented by Simons (1983, pp. 487–492), Kinzelbach (1986, pp. 68–75), and Bourke (1987, pp. 143–150). The theory of contouring is presented by Davis (1986, pp. 353–376).

It is often desirable to view drawdown cones of depression and contaminant concentration plumes in three-dimensional form. Surfaces may be displayed at various angles and rotations to add greatly to the realism of a picture. A 3-D surface plot microcomputer program is listed by Fowler (1984, pp. 292–296) and the theory of 3-D plots is presented by Hearn and Baker (1983, pp. 219–278).

Goldstein (1984, p.61) presents programs for saving on diskette and recalling to the screen from a diskette entire screen images. BASIC initially allows three open diskette files at a time. This number can be increased by giving the appropriate command when starting BASIC (see Goldstein, 1984, p. 113).

BASIC programs for tracing streamlines and contaminant particle-tracks with dispersion on the screen are listed by Bear and Verruijt (1987, pp. 336–343).

Trend Graph

It is often desirable to find the best-fit trend line through pumping test time-drawdown semilog data points so that values of aquifer transmissivity and storativity may be determined with predicted values of drawdown along the line. The regression least-squares method uses differential calculus to arrive at the slope and the Y intercept of the best-fit line using the following equations (see Davis, 1986, p. 180):

$$c_1 = [\sum_{i=1}^{n} X_i Y_i - (\sum_{i=1}^{n} X_i \sum_{i=1}^{n} Y_i)/n]/[\sum_{i=1}^{n} X^2_i - (\sum_{i=1}^{n} X_i)^2/n]$$

(53)

$$c_0 = \sum_{i=1}^{n} Y_i/n - c_1 \sum_{i=1}^{n} X_i/n$$ (54)

where

c_1 = slope of best-fit line as a decimal

c_0 = zero drawdown-intercept of best-fit line in min

Y_i = specified value of drawdown in ft

X_i = specified value of time in minutes

n = number of specified X_i and Y_i values

Values of drawdown for specified times along the best-fit line may be calculated knowing c_1 and c_0 with the following equation of a line (see Davis, 1986, p. 179):

$$Y_t = c_0 + c_1 X_t$$ (55)

where

Y_t = calculated value of drawdown along best-fit line in ft

X_t = specified value of time along best-fit line in minutes

Data for any two points on the best-fit line may be substituted in the following equation to calculate aquifer transmissivity (see Cooper and Jacob, 1946, pp. 526–534):

$$T = 264Q\log(t_2/t_1)/(s_2 - s_1) \qquad (56)$$

where

T = aquifer transmissivity in gpd/ft

Q = production well discharge rate in gpm

s_1 = drawdown at selected point 1 in ft

s_2 = drawdown at selected point 2 in ft

t_1 = time at selected point 1 in minutes

t_2 = time at selected point 2 in minutes

The calculated value of aquifer transmissivity and data for any point on the best-fit line may be substituted in the following equation to calculate aquifer storativity (see Cooper and Jacob, 1946, pp. 526–534):

$$S = Tt/[2693r^2e^{\ln(1/u)}] \qquad (57)$$

where

$$\ln(1/u) = sT/(114.6Q) + 0.5772 \qquad (58)$$

s = drawdown in ft

S = aquifer storativity as a decimal

t = time in minutes

r = distance from production well in ft

e = 2.71828183

Lines frequently connect data points in arithmetic time-drawdown or time-concentration graphs. Continuous curves may be drawn through data points by using Equation 22 (Lagrange interpolation) or a spline program (see Fowler, 1984, pp. 301–305).

Surface Interpolation

Programs that generate contour maps and 3-D plots require XYZ data in a regularly spaced (gridded) format. This format may be created by interpolating between grid points and irregular and scattered control data points with triangulation or kriging techniques (see Davis, 1986, pp. 353–405). Interpolation is based on inverse distance squared weighted average methods. In the triangulation technique, distances between grid points and control data points are calculated with the following Pythagorean equation (Davis, 1986, p. 367):

$$D_c = [(X_c - X_g)^2 + (Y_c - Y_g)^2]^{1/2} \tag{59}$$

where

D_c = distance between grid point g and control data point c in ft

X_c = control data point X-coordinate in ft

X_g = grid point X-coordinate in ft

Y_c = control data point Y-coordinate in ft

Y_g = grid point Y-coordinate in ft

Distances are sorted in ascending order and the six control data points closest to the grid point of interest are identified. The grid point surface value is estimated using the surface values for these control data points and the following weighting function equation (see Davis, 1986, pp. 367–371):

$$RV_g = \sum_{c=1}^{6} (WF_c/WT)(RV_c/D_c)/[\ \sum_{c=1}^{6} (WF_c/WT)(1/D_c)] \tag{60}$$

where

$$WF_c = (1 - D_c/D_6)^2/(D_c/D_6) \tag{61}$$

$$WT = \sum_{c=1}^{6} WF_c \tag{62}$$

RV_g = estimated surface value at grid point g in ft or mg/L

RV_c = surface value at control data point c in ft or mg/L

D_6 = distance between grid point g and farthest control data point in ft

WF_c = inverse distance-squared weighting function for control data point c (weightings vary according to the distances between the grid node being estimated and the control data points)

The procedure is repeated for each grid point. Kriging procedures are outlined by Davis (1986, pp. 353–405), Marsily (1986, pp. 284–337), and Olea (1975, p.137).

The contouring algorithm in GWGRAF consists of the detection of each contour line intersecting the four sides of a given grid square and drawing contour lines through the grid-square block. Two grid-node values along a grid-square side are compared with the value of a contour. If the contour value lies between two of the grid-node values, the point location of the crossing is calculated by linear interpolation using the following equation (Davis, 1986, p.147):

$$C_i = [(C_2 - C_1)/(S_2 - S_1)](S_i - S_1) + C_1 \qquad (63)$$

where

S_i = value of contour in ft or mg/L

S_1 = grid-node 1 surface value in ft or mg/L

S_2 = grid-node 2 surface value in ft or mg/L

C_1 = grid-node 1 X or Y coordinate in ft or mg/L

C_2 = grid-node 2 X or Y coordinate in ft or mg/L

C_i = X or Y coordinate of contour point location in ft or mg/L

This procedure is repeated for all of the grid-square sides. The grid square is then subdivided into smaller squares, and intermediate point locations of the contour line within the grid-square block are interpolated with Equation 63. Screen pixels are turned on at point locations of the contour. Interpolation through both I and J directions is performed to ensure that contours parallel to both axes are considered. Selection of smaller squares is based on the desired density of intermediate point locations of the contour. The small square size relates to the pixel (dot) size on the screen; with sufficiently fine small square sizes, dots representing point contour locations merge together to look like continuous curves.

Appendix A
WELFUN Source Code

The source code for the WELFUN IBM PC BASICA microcomputer program is listed in this appendix. Program algorithms are documented as REM statements in the listing to assist the reader in understanding the code. Many statements in the program could be restructured or eliminated to produce more efficient coding, but they have been intentionally written in their present form to provide clear explanations of processing steps.

```
10 CLS:CLEAR:KEY OFF
20 DIM UAT(19),WAT(19,8),UBT(33),WBT(33,8),BETA1(8),B1DIF(7)
30 AF$="##.####^^^^"
40 FLAGP=1
50 PRINT"Program: WELFUN"
60 PRINT"Version: IBM/PC BASICA 2.1"
70 PRINT"          Copyright 1988 Lewis Publishers, Inc."
80 PRINT"Purpose: Calculate values of Well Functions"
90 PRINT"Author : William C. Walton"
100 PRINT"Date    : April, 1988"
110 PRINT:PRINT"Do you have a printer?"
120 INPUT"Enter 1 for yes or 2 for no";AI
130 IF AI <> 1 AND AI <> 2 THEN 110
140 PRINT:PRINT"WELL FUNCTION OPTIONS:"
150 PRINT:PRINT"Enter 1 for W(u)"
160 PRINT"Enter 2 for W(u,r/B)"
170 PRINT"Enter 3 for W(uA, uB, Beta)"
180 PRINT"Enter 4 for Wfp(uf,r/mb,b,c)"
190 INPUT"Enter 5 for W(u,rPv/Ph^.5/m,L/m,d/m,Lo/m,do/m)";WFO
200 IF WFO <> 1 AND WFO <> 2 AND WFO <> 3 AND WFO <> 4 AND WFO <> 5 THEN 150
210 IF WFO=5 THEN 280
220 PRINT:PRINT"WELL FUNCTION COMBINATION OPTIONS:"
230 PRINT:PRINT"Enter 6 to combine any of the first 4 Well Functions"
240 PRINT"listed above with partial penetration impact Well"
250 PRINT"Function W(u,rPv/Ph^.5/m,L/m,d/m,Lo/m,do/m)"
260 INPUT"or 7 for no combination of Well Functions";WFC
270 IF WFC <> 6 AND WFC <> 7 THEN 230
280 IF FLAGP>1 THEN 340
290 IF AI=2 THEN 330
300 PRINT:PRINT"Turn on the printer or else an error will occur"
310 PRINT:PRINT"Press any key to continue"
320 A$=INKEY$:IF A$="" THEN 320
330 FLAGP=FLAGP+1
340 PRINT:PRINT"DATA BASE ENTRY:"
350 IF AI=1 THEN LPRINT
360 IF AI=1 THEN LPRINT"DATA BASE:"
```

```
370 IF WFO <> 1 THEN 1600
380 'DATA BASE FOR W(U)
390 PRINT:INPUT"u=";U
400 IF AI=1 THEN LPRINT
410 IF AI=1 THEN LPRINT"u=" USING AF$;U
420 GOSUB 2830
430 PRINT:PRINT"COMPUTATION RESULTS:"
440 IF AI=1 THEN LPRINT
450 IF AI=1 THEN LPRINT"COMPUTATION RESULTS:"
460 PRINT:PRINT"W(u)=" USING AF$;W
470 IF AI=1 THEN LPRINT
480 IF AI=1 THEN LPRINT"W(u)=" USING AF$;W
490 IF WFC=7 THEN 510
500 GOSUB 580
510 PRINT:PRINT"Enter 1 for another value of W(u)"
520 PRINT"Enter 2 for another Well Function"
530 INPUT"Enter 3 to end program";WFR
540 IF WFR <> 1 AND WFR <> 2 AND WFR <> 3 THEN 510
550 IF WFR=1 THEN 210
560 IF WFR=2 THEN 150
570 IF WFR=3 THEN 5310
580 PRINT:PRINT"PARTIAL PENETRATION DATA BASE BASE ENTRY:"
590 IF AI=1 THEN LPRINT
600 IF AI=1 THEN LPRINT"PARTIAL PENETRATION DATA BASE:"
610 Z=0
620 IF AI=1 THEN LPRINT
630 PRINT
640 PRINT"Enter 1 for production well function"
650 INPUT"Enter 2 for observation well function";PPW
660 IF PPW <> 1 AND PPW <> 2 THEN 640
670 INPUT"Aquifer horiz. hydraulic conductivity (gpd/sq ft)=";PH
680 IF AI=2 THEN 700
690 LPRINT"Aquifer horiz. hydraulic conductivity (gpd/sq ft)=" USING AF$;PH
700 INPUT"Aquifer vert. hydraulic conductivity (gpd/sq ft)=";PV
710 IF AI=2 THEN 730
720 LPRINT"Aquifer vert. hydraulic conductivity (gpd/sq ft)=" USING AF$;PV
730 IF PW=1 THEN 780
740 INPUT"Radial distance to well in ft=";RD
750 IF AI=2 THEN 770
760 LPRINT"Radial distance to well in ft=" USING AF$;RD
770 GOTO 810
780 INPUT"Production well radius in ft=";RD
```

```
790 IF AI=2 THEN 810
800 LPRINT"Production well radius in ft=" USING AF$;RD
810 INPUT"Aquifer thickness in ft=";MA
820 IF AI=2 THEN 840
830 LPRINT"Aquifer thickness in ft=" USING AF$;MA
840 INPUT"Distance from aquifer top to bottom of prod. well (ft)=";LP
850 IF AI*2 THEN 870
860 LPRINT"Dist. from aquifer top to bottom of prod. well (ft)=" USING AF$;LP
870 INPUT"Dist. from aquifer top to top of prod. well screen (ft)=";DP
880 IF AI=2 THEN 900
890 LPRINT"Dist. from aquifer top to top of prod. well screen(ft)" USING AF$;DP
900 IF PPW=1 THEN 980
910 INPUT"Distance from aquifer top to bottom of obs. well (ft)=";LL
920 IF AI=2 THEN 940
930 LPRINT"Dist. from aquifer top to bottom of obs. well (ft)=" USING AF$;LL
940 INPUT"Dist. from aquifer top to top of obs. well screen (ft)=";DD
950 IF AI=2 THEN 970
960 LPRINT"Dist. from aquifer top to top of obs. well screen(ft)=" USING AF$;DD
970 GOTO 1010
980 LL=(LP+DP)/2
990 DD=LP-1
1000 'SUBROUTINE TO CALCULATE W(U,R(PV/PH)^.5/M,L/M,D/M,LO/M,DO/M)
1010 PRINT:PRINT"COMPUTATIONS ARE IN PROGRESS":PRINT
1020 FOR K=1 TO 50
1030 S=K*3.1416*RD*(PV/PH)^.5/MA
1040 GOSUB 2950
1050 AA=SIN(K*3.1416*(LP/MA))-SIN(K*3.1416*(DP/MA))
1060 BB=SIN(K*3.1416*(LL/MA))-SIN(K*3.1416*(DD/MA))
1070 Z=Z+1/K^2*(AA*BB*WS)
1080 NEXT K
1090 WUPAR=2*MA^2*Z/(3.1416^2*(LP-DP)*(LL-DD))
1100 IF WFO=2 THEN W=WS1
1110 IF WFO=3 AND WUB=1 THEN W=WA
1120 IF WFO=3 AND WUB=2 THEN W=WB
1130 IF WFO=4 THEN W=WF
1140 WUPT=W+WUPAR
1150 IF AI=1 THEN LPRINT
1160 PRINT
1170 IF WFO=5 THEN 1200
1180 IF AI=1 THEN LPRINT"COMPUTATION RESULTS:"
1190 IF AI=1 THEN LPRINT
1200 IF WFO <> 1 THEN 1280
```

```
1210 PRINT"W(u)+W(u,Pv/Ph^.5/m,L/m,d/m,Lo/m,do/m)=" USING AF$;WUPT
1220 IF AI=2 THEN 1240
1230 LPRINT"W(u)+W(u,rPv/Ph^.5/m,L,/m,d/m,Lo/m,do/m)=" USING AF$;WUPT
1240 PRINT"W(u,rPv/Ph^.5/m,L/m,d/m,Lo/m,do/m)=" USING AF$;WUPAR
1250 IF AI=2 THEN 1270
1260 LPRINT"W(u,rPv/Ph^.5/m,L/m,d/m,Lo/m,do/m)=" USING AF$;WUPAR
1270 RETURN
1280 IF WFO <> 2 THEN 1360
1290 PRINT"W(u,r/B,rPv/Ph^.5/m,L/m,d/m,Lo/m,do/m)=" USING AF$;WUPT
1300 IF AI=2 THEN 1320
1310 LPRINT"W(u,r/B,rPv/Ph^.5/m,L,/m,d/m,Lo/m,do/m)=" USING AF$;WUPT
1320 PRINT"W(u,rPv/Ph^.5/m,L/m,d/m,Lo/m,do/m)=" USING AF$;WUPAR
1330 IF AI=2 THEN 1350
1340 LPRINT"W(u,rPv/Ph^.5/m,L/m,d/m,Lo/m,do/m)=" USING AF$;WUPAR
1350 RETURN
1360 IF WFO <> 3 THEN 1520
1370 IF WUB=2 THEN 1450
1380 PRINT"W(uA,Beta,rPv/Ph^.5/m,L/m,d/m,Lo/m,do/m)=" USING AF$;WUPT
1390 IF AI=2 THEN 1410
1400 LPRINT"W(uA,Beta,rPv/Ph^.5/m,L,/m,d/m,Lo/m,do/m)=" USING AF$;WUPT
1410 PRINT"W(u,rPv/Ph^.5/m,L/m,d/m,Lo/m,do/m)=" USING AF$;WUPAR
1420 IF AI=2 THEN 1440
1430 LPRINT"W(u,rPv/Ph^.5/m,L/m,d/m,Lo/m,do/m)=" USING AF$;WUPAR
1440 IF WUB=1 THEN 1510
1450 PRINT"W(uB,Beta,rPv/Ph^.5/m,L/m,d/m,Lo/m,do/m)=" USING AF$;WUPT
1460 IF AI=2 THEN 1480
1470 LPRINT"W(uB,Beta,rPv/Ph^.5/m,L,/m,d/m,Lo/m,do/m)=" USING AF$;WUPT
1480 PRINT"W(u,rPv/Ph^.5/m,L/m,d/m,Lo/m,do/m)=" USING AF$;WUPAR
1490 IF AI=2 THEN 1510
1500 LPRINT"W(u,rPv/Ph^.5/m,L/m,d/m,Lo/m,do/m)=" USING AF$;WUPAR
1510 RETURN
1520 IF WFO <> 4 THEN 1590
1530 PRINT"W(uf,r/mb,b,c,rPv/Ph^.5/m,L/m,d/m,Lo/m,do/m)=" USING AF$;WUPT
1540 IF AI=2 THEN 1560
1550 LPRINT"W(uf,r/mb,b,c,rPv/Ph^.5/m,L,/m,d/m,Lo/m,do/m)=" USING AF$;WUPT
1560 PRINT"W(u,rPv/Ph^.5/m,L/m,d/m,Lo/m,do/m)=" USING AF$;WUPAR
1570 IF AI=2 THEN 1590
1580 LPRINT"W(u,rPv/Ph^.5/m,L/m,d/m,Lo/m,do/m)=" USING AF$;WUPAR
1590 RETURN
1600 IF WFO <> 2 THEN 1860
1610 'DATA BASE FOR W(U,R/B)
1620 IF AI=1 THEN LPRINT
```

```
1630 PRINT
1640 FLAG=1
1650 INPUT"u=";U
1660 INPUT"r/B=";S
1670 IF AI=1 THEN LPRINT"u=" USING AF$;U
1680 IF AI=1 THEN LPRINT"r/B=" USING AF$;S
1690 GOSUB 2840
1700 GOSUB 2950
1710 PRINT:PRINT"COMPUTATION RESULTS:"
1720 IF AI=1 THEN LPRINT
1730 IF AI=1 THEN LPRINT"COMPUTATION RESULTS:"
1740 PRINT:PRINT"W(u,r/B)=" USING AF$;WS
1750 IF AI=1 THEN LPRINT
1760 IF AI=1 THEN LPRINT"W(u,r/B)=" USING AF$;WS
1770 IF WFC=7 THEN 1790
1780 GOSUB 580
1790 PRINT:PRINT"Enter 1 for another value of W(u,r/B)"
1800 PRINT"Enter 2 for another Well Function"
1810 INPUT"Enter 3 to end program";WFR
1820 IF WFR <> 1 AND WFR <> 2 AND WFR <> 3 THEN 1790
1830 IF WFR=1 THEN 210
1840 IF WFR=2 THEN 150
1850 IF WFR=3 THEN 5310
1860 IF WFO <> 3 THEN 2330
1870 'DATA BASE FOR W(UA,UB,BETA)
1880 IF FLAGA>1 THEN  1900
1890 FLAGA=1
1900 PRINT
1910 IF AI=1 THEN LPRINT
1920 PRINT"Enter 1 for W(uA,Beta)"
1930 INPUT"Enter 2 for W(uB,Beta)";WUB
1940 IF WUB <> 1 AND WUB <> 2 THEN 1920
1950 PRINT
1960 IF WUB=2 THEN 2010
1970 INPUT"uA=";UA
1980 U=UA
1990 UA=1/UA
2000 GOTO 2040
2010 INPUT"uB=";UB
2020 U=UB
2030 UB=1/UB
2040 INPUT"Beta=";BETA
```

```
2050 IF WUB=2 THEN 2080
2060 IF AI=1 THEN LPRINT"uA=" USING AF$;U
2070 GOTO 2090
2080 IF AI=1 THEN LPRINT"uB=" USING AF$;U
2090 IF AI=1 THEN LPRINT"Beta=" USING AF$;BETA
2100 GOSUB 3610
2110 FLAGA=FLAGA+1
2120 PRINT:PRINT"COMPUTATION RESULTS:"
2130 IF AI=1 THEN LPRINT
2140 IF AI=1 THEN LPRINT"COMPUTATION RESULTS:"
2150 IF WUB=2 THEN 2200
2160 PRINT:PRINT"W(uA,Beta)=" USING AF$;WA
2170 IF AI=1 THEN LPRINT
2180 IF AI=1 THEN LPRINT"W(uA,Beta)=" USING AF$;WA
2190 GOTO 2230
2200 PRINT:PRINT"W(uB,Beta)=" USING AF$;WB
2210 IF AI=1 THEN LPRINT
2220 IF AI=1 THEN LPRINT"W(uB,Beta)=" USING AF$;WB
2230 IF WFC=7 THEN 2260
2240 GOSUB 2840
2250 GOSUB 580
2260 PRINT:PRINT"Enter 1 for another value of W(uA,Beta) or W(uB,beta)"
2270 PRINT"Enter 2 for another Well Function"
2280 INPUT"Enter 3 to end program";WFR
2290 IF WFR <> 1 AND WFR <> 2 AND WFR <> 3 THEN 2260
2300 IF WFR=1 THEN 210
2310 IF WFR=2 THEN 150
2320 IF WFR=3 THEN 5310
2330 IF WFO <> 4 THEN 2640
2340 'DATA BASE FOR WFP(UF,R/MB,B,C)
2350 IF AI=1 THEN LPRINT
2360 PRINT
2370 INPUT"uf=";UF
2380 INPUT"r/mb=";RMB
2390 INPUT"b=";B
2400 INPUT"c=";C
2410 IF AI=2 THEN 2460
2420 LPRINT"uf=" USING AF$;UF
2430 LPRINT"r/mb=" USING AF$;RMB
2440 LPRINT"b=" USING AF$;B
2450 LPRINT"c=" USING AF$;C
2460 GOSUB 5130
```

```
2470 PRINT:PRINT"COMPUTATION RESULTS:"
2480 IF AI=1 THEN LPRINT
2490 IF AI=1 THEN LPRINT"COMPUTATION RESULTS:"
2500 PRINT:PRINT"Wfp(uf,r/mb,b,c)=" USING AF$;WF
2510 IF AI=1 THEN LPRINT
2520 IF AI=1 THEN LPRINT"Wfp(ub,r/mb,b,c)=" USING AF$;WF
2530 IF WFC=7 THEN 2560
2540 GOSUB 2840
2550 GOSUB 580
2560 PRINT:PRINT"Enter 1 for another value of Wfp(uf,r/mb,b,c)
2570 PRINT"Enter 2 for another Well Function"
2580 INPUT"Enter 3 to end program";WFR
2590 IF WFR <> 1 AND WFR <> 2 AND WFR <> 3 THEN 2560
2600 IF WFR=1 THEN 210
2610 IF WFR=2 THEN 150
2620 IF WFR=3 THEN 5310
2630 IF AI=1 THEN LPRINT
2640 PRINT
2650 'DATA BASE FOR W(U,RPV/PH^.5/M,L/M,D/M,LO/M,DO/M)
2660 INPUT"u=";U
2670 IF AI=1 THEN LPRINT"u=" USING AF$;U
2680 GOSUB 580
2690 PRINT:PRINT"COMPUTATION RESULTS:":PRINT
2700 IF AI=1 THEN LPRINT
2710 IF AI=1 THEN LPRINT"COMPUTATION RESULTS:"
2720 PRINT"W(u,rPv/Ph^.5/m,L/m,d/m,Lo/m,do/m)=" USING AF$;WUPAR
2730 IF AI=2 THEN 2750
2740 LPRINT"W(u,rPv/Ph^.5/m,L/m,d/m,Lo/m,do/m)=" USING AF$;WUPAR
2750 PRINT
2760 PRINT"Enter 1 for another value of W(u,rPv/Ph^.5/m,L/m,d/m,Lo/m,do/m)"
2770 PRINT"Enter 2 for another Well Function"
2780 INPUT"Enter 3 to end program";WPP
2790 IF WPP <> 1 AND WPP <> 2 AND WPP <> 3 THEN 2760
2800 IF WPP=1 THEN 210
2810 IF WPP=2 THEN 150
2820 IF WPP=3 THEN 5310
2830 'SUBROUTINE TO COMPUTE W(U) USING POLYNOMIAL APPROXIMATIONS
2840 A=U^2:B=U^3:C=U^4:D=U^5
2850 IF U>10 THEN W=0
2860 IF U>10 THEN 2930
2870 IF U>1 THEN 2900
2880 W=-LOG(U)-.5772156600000006#+.9999919300000007#*U-.24991055#*A+.0551997*B-9
.76004E-03*C+1.07857E-03*D
```

```
2890 GOTO 2930
2900 W=C+8.573328740100012#*B+18.059016973#*A+8.634760892499999#*U+.26777373430
0006#
?910 W=W/(C+9.573322345400008#*B+25.6329561486#*A+21.0996530827#*U+3.9584969228
0006#)
2920 W=W/(U*EXP(U))
2930 RETURN
2940 'SUBROUTINE TO COMPUTE W(U,R/B) USING POLYNOMIAL APPROXIMATIONS
2950 IF U>10 THEN WS=0:GOTO 3340
2960 IF S>2 THEN 3240
2970 L=S/3.75
2980 V=1+3.5156229#*L^2+3.0899424#*L^4+1.2067492#*L^6+.265973*L^8+.0360768*L^10
2990 V=V+.0045813*L^12
3000 F=S/2
3010 H=-LOG(F)*V-.5772156600000009#+.422784*F^2+.23069756#*F^4+.0348859*F^6
3020 H=H+2.62698E-03*F^8+.0001075*F^10+.0000074*F^12
3030 IF S=0 THEN WS=W:GOTO 3340
3040 N=S^2/(4*U)
3050 IF N>5 THEN WS=2*H:GOTO 3340
3060 IF U<=.9 THEN 3090
3070 A=U+.5858:B=U+3.414:C=S*S/4
3080 WS=1.5637*EXP(-A-C/A)/A+4.54*EXP(-B-C/B)/B:GOTO 3340
3090 IF U<.05 THEN 3130
3100 IF U>S/2 THEN 3120
3110 C=-(1.75*U)^-.448*S:WS=2*H-4.8*10^C:GOTO 3340
3120 WS=W-(S/(4.7*U^.6))^2:GOTO 3340
3130 IF U>.01 AND S<.1 THEN 3120
3140 IF N<1 THEN 3210
3150 WN=N^4+8.573328740100025#*N^3+18.059016973#*N^2+8.634760892499999#*N
3160 WN=WN+.26777373430000001#
3170 WNP=N^4+9.573322345400009#*N^3+25.6329561486#*N^2+21.0996530827#*N
3180 WNP=WNP+3.958496922800012#
3190 WN=WN/WNP
3200 WN=WN/(N*EXP(N)):GOTO 3230
3210 WN=-LOG(N)+.99999193000000017#*N-.5772156600000009#-.24991055#*N^2
3220 WN=WN+.0551997*N^3-9.76004E-03*N^4+1.07857E-03*N^5
3230 WS=2*H-WN*V:GOTO 3340
3240 M=-((S-2*U)/(2*U^.5))
3250 IF M<0 THEN M=ABS(M):GOSUB 3270:GOTO 3330
3260 GOSUB 3270:GOTO 3320
3270 NP=1+.0705230784#*M+.0422820123#*M^2+9.270527200000026D-03*M^3
3280 NP=NP+1.52014E-04*M^4+2.76567E-04*M^5+4.30638E-05*M^6
```

```
3290 IF NP>100! THEN 3310
3300 N=1/NP^16:RETURN
3310 N=0!:RETURN
3320 WS=(3.1416/(2*S))^.5*EXP(-S)*N:GOTO 3340
3330 WS=(3.1416/(2*S))^.5*EXP(-S)*(2-N)
3340 IF FLAG>1 THEN 3360
3350 WS1=WS
3360 FLAG=FLAG+1
3370 RETURN
3380 IF WFO <> 2 THEN 2090
3390 INPUT"u=";U
3400 INPUT"r/B=";S
3410 IF AI=2 THEN 3440
3420 LPRINT"u="USING AF$;U
3430 LPRINT"r/B="USING AF$;S
3440 GOSUB 2830
3450 GOSUB 2940
3460 W1=WS:PRINT:PRINT"COMPUTATION RESULTS:"
3470 IF AI=1 THEN LPRINT
3480 IF AI=1 THEN LPRINT"COMPUTATION RESULTS:"
3490 PRINT:PRINT"W(u,r/B)=" USING AF$;WS
3500 IF AI=1 THEN LPRINT
3510 IF AI=1 THEN LPRINT"W(u,r/B)=" USING AF$;WS
3520 IF WFC=6 THEN 580
3530 IF WFC=6 THEN 1000
3540 PRINT:PRINT"Enter 1 for another value of W(u,r/B)"
3550 PRINT"Enter 2 for another Well Function"
3560 INPUT"Enter 3 to end program";WFR
3570 IF WFR <> 1 AND WFR <> 2 AND WFR <> 3 THEN 3540
3580 IF WFR=1 THEN 210
3590 IF WFR=2 THEN 140
3600 IF WFR=3 THEN 5310
3610 PRINT:PRINT"COMPUTATIONS ARE IN PROGRESS":PRINT
3620 'SUBROUTINE TO CALCULATE W(UA,UB,BETA) USING TABLE INTERPOLATION
3630 IF FLAGA>1 THEN 4420
3640 FOR J=1 TO 19
3650 READ UAT(J)
3660 NEXT J
3670 'VALUES OF 1/UA IN TABLE
3680 DATA 4E-1,8E-1,1.4,2.4,4,8,1.4E1,2.4E1,4E1,8E1,1.4E2,2.4E2
3690 DATA 4E2,8E2,1.4E3,2.4E3,4E3,8E3,1.4E4
3700 FOR K=1 TO 8
```

```
3710 FOR J=1 TO 19
3720 READ WAT(J,K)
3730 NEXT J
3740 NEXT K
3750 'VALUES OF W(1/UA,BETA) IN TABLE
3760 DATA 2.48E-2,1.45E-1,3.58E-1,6.62E-1,1.02,1.57,2.05,2.52
3770 DATA 2.97,3.56,4.01,4.42,4.77,5.16,5.40,5.54,5.59,5.62,5.62
3780 DATA 2.41E-2,1.40E-1,3.45E-1,6.33E-1,9.63E-1,1.46,1.88,2.27
3790 DATA 2.61,3.00,3.23,3.37,3.43,3.45,3.46,3.46,3.46,3.46,3.46
3800 DATA 2.30E-2,1.31E-1,3.18E-1,5.70E-1,8.49E-1,1.23,1.51,1.73
3810 DATA 1.85,1.92,1.93,1.94,1.94,1.94,1.94,1.94,1.94,1.94,1.94
3820 DATA 2.14E-2,1.19E-1,2.79E-1,4.83E-1,6.88E-1,9.18E-1,1.03,1.07
3830 DATA 1.08,1.08,1.08,1.08,1.08,1.08,1.08,1.08,1.09,1.09,1.09
3840 DATA 1.88E-2,9.88E-2,2.17E-1,3.43E-1,4.38E-1,4.97E-1,5.07E-1
3850 DATA 5.07E-1,5.07E-1,5.07E-1,5.07E-1,5.07E-1,5.07E-1,5.07E-1
3860 DATA 5.07E-1,5.08E-1,5.08E-1,5.08E-1,5.08E-1
3870 DATA 1.70E-2,8.49E-2,1.75E-1,2.56E-1,3.00E-1,3.17E-1,3.17E-1
3880 DATA 3.17E-1,3.17E-1,3.17E-1,3.17E-1,3.17E-1,3.17E-1,3.17E-1
3890 DATA 3.17E-1,3.17E-1,3.18E-1,3.18E-1,3.18E-1
3900 DATA 1.38E-2,6.03E-2,1.07E-1,1.33E-1,1.40E-1,1.41E-1,1.41E-1
3910 DATA 1.41E-1,1.41E-1,1.41E-1,1.41E-1,1.41E-1,1.41E-1,1.41E-1
3920 DATA 1.41E-1,1.41E-1,1.41E-1,1.42E-1,1.42E-1
3930 DATA 9.33E-3,3.17E-2,4.45E-2,4.76E-2,4.78E-2,4.78E-2,4.78E-2
3940 DATA 4.78E-2,4.78E-2,4.78E-2,4.78E-2,4.78E-2,4.78E-2,4.78E-2
3950 DATA 4.78E-2,4.78E-2,4.78E-2,4.79E-2,4.79E-2
3960 FOR J=1 TO 33
3970 READ UBT(J)
3980 NEXT J
3990 'VALUES OF 1/UB IN TABLE
4000 DATA 4.0E-4,8.0E-4,1.4E-3,2.4E-3,4.0E-3,8.0E-3,1.4E-2,2.4E-2
4010 DATA 4.0E-2,8.0E-2,1.4E-1,2.4E-1,4.0E-1,8.0E-1,1.4,2.4,4.0,8.0
4020 DATA 1.4E1,2.4E1,4.0E1,8.0E1,1.4E2,2.4E2,4.0E2,8.0E2,1.4E3,2.4E3
4030 DATA 4.0E3,8.9E3,1.4E4,2.4E4,4.0E4
4040 FOR K=1 TO 8
4050 FOR J=1 TO 33
4060 READ WBT(J,K)
4070 NEXT J
4080 NEXT K
4090 'VALUES OF W(1/UB,BETA) IN TABLE
4100 DATA 5.62,5.62,5.62,5.62,5.62,5.62,5.62,5.62,5.62,5.62,5.62,5.62
4110 DATA 5.62,5.62,5.63,5.63,5.63,5.64,5.65,5.67,5.70,5.76,5.85,5.99
4120 DATA 6.16,6.47,6.67,7.21,7.72,8.41,8.97,9.51,1.94E1
```

```
4130 DATA 3.46,3.46,3.46,3.46,3.46,3.46,3.46,3.46,3.46,3.46,3.46,3.46
4140 DATA 3.46,3.46,3.47,3.49,3.51,3.56,3.63,3.74,3.90,4.22,4.58,5.00
4150 DATA 5.46,6.11,6.67,7.21,7.72,8.41,8.97,9.51,1.94E1
4160 DATA 1.94,1.94,1.94,1.94,1.94,1.94,1.94,1.94,1.94,1.94,1.94,1.95
4170 DATA 1.96,1.98,2.01,2.06,2.13,2.31,2.55,2.86,3.24,3.85,4.38,4.91
4180 DATA 5.42,6.11,6.67,7.21,7.72,8.41,8.97,9.51,1.94E1
4190 DATA 1.09,1.09,1.09,1.09,1.09,1.09,1.09,1.09,1.09,1.09,1.10,1.11
4200 DATA 1.13,1.18,1.24,1.35,1.50,1.85,2.23,2.68,3.15,3.82,4.37,4.91
4210 DATA 5.42,6.11,6.67,7.21,7.72,8.41,8.97,9.51,1.94E1
4220 DATA 5.08E-1,5.08E-1,5.08E-1,5.08E-1,5.08E-1,5.09E-1,5.10E-1,5.12E-1
4230 DATA 5.16E-1,5.24E-1,5.37E-1,5.57E-1,5.89E-1,5.67E-1,7.80E-1,9.54E-1
4240 DATA 1.20,1.68,2.15,2.65,3.14,3.82,4.37,4.91,5.42,6.11,6.67,7.21,7.72
4250 DATA 8.41,8.97,9.51,1.94E1
4260 DATA 3.18E-1,3.18E-1,3.18E-1,3.18E-1,3.18E-1,3.19E-1,3.21E-1,3.23E-1
4270 DATA 3.27E-1,3.37E-1,3.50E-1,3.74E-1,4.12E-1,5.06E-1,6.42E-1,8.50E-1
4280 DATA 1.13,1.65,2.14,2.65,3.14,3.82,4.37,4.91,5.42,6.11,6.67,7.21,7.72
4290 DATA 8.41,8.97,9.51,1.94E1
4300 DATA 1.42E-1,1.42E-1,1.42E-1,1.42E-1,1.42E-1,1.43E-1,1.45E-1,1.47E-1
4310 DATA 1.52E-1,1.62E-1,1.78E-1,2.05E-1,2.48E-1,3.57E-1,5.17E-1,7.63E-1
4320 DATA 1.08,1.63,2.14,2.64,3.14,3.82,4.37,4.91,5.42,6.11,6.67,7.21,7.72
4330 DATA 8.41,8.97,9.51,1.94E1
4340 DATA 4.79E-2,4.80E-2,4.81E-2,4.84E-2,4.88E-2,4.96E-2,5.09E-2,5.32E-2
4350 DATA 5.68E-2,6.61E-2,8.06E-2,1.06E-1,1.49E-1,2.66E-1,4.45E-1,7.18E-1
4360 DATA 1.06,1.63,2.14,2.64,3.14,3.82,4.37,4.91,5.42,6.11,6.67,7.21,7.72
4370 DATA 8.41,8.97,9.51,1.94E1
4380 FOR K=1 TO 8
4390 READ BETA1(K)
4400 NEXT K
4410 DATA 1E-3,1E-2,6E-2,2E-1,6E-1,1,2,4
4420 IF BETA >4! OR BETA <.001 THEN 4480
4430 IF WUB=2 THEN 4460
4440 IF UA<.4 OR UA>14000 THEN 4480
4450 GOTO 4570
4460 IF UB<.0004 OR UB>40000! THEN 4480
4470 GOTO 4570
4480 PRINT:PRINT"DATA BASE EXCEEDS PROGRAM LIMITS"
4490 PRINT"(BETA>=1E-3 AND BETA<=4.0,1/UA>=4E-1 AND 1/UA<=1.4E4,"
4500 PRINT"1/UB>=4E-4 AND 1/UB<=4E4) AND MUST BE REVISED."
4510 PRINT:PRINT"Press any key to continue"
4520 T$=INKEY$:IF T$="" THEN 4520
4530 PRINT
4540 FLAGA=FLAGA+1
```

```
4550 GOTO 210
4560 'IDENTIFYING CLOSEST BETA VALUES
4570 FOR J=1 TO 7
4580 FLAG3=J:FLAG4=J+1
4590 IF BETA>=BETA1(J) AND BETA<=BETA1(J+1) THEN 4610
4600 NEXT J
4610 BA1=BETA1(FLAG3):BA2=BETA1(FLAG4)
4620 IF WUB <> 1 THEN 4860
4630 'IDENTIFYING CLOSEST 1/UA VALUES
4640 FOR J=1 TO 18
4650 FLAG1=J:FLAG2=J+1
4660 IF UA>=UAT(J) AND UA<=UAT(J+1) THEN 4680
4670 NEXT J
4680 UA1=UAT(FLAG1):UA2=UAT(FLAG2)
4690 'INTERPOLATION OF 1/UA AND W(1/UA,BETA) VALUES
4700 IF BETA=BETA1(8) AND UA=UAT(19) THEN WA=WAT(19,8)
4710 IF BETA=BETA1(8) AND UA=UAT(19) THEN 5100
4720 IF BETA=BETA1(8) THEN 4800
4730 IF UA=UAT(19) THEN 4830
4740 WA1=WAT(FLAG1,FLAG3):WA2=WAT(FLAG1,FLAG4)
4750 WA3=WAT(FLAG2,FLAG3):WA4=WAT(FLAG2,FLAG4)
4760 WA5=WA1-((BETA-BA1)/(BA2-BA1))*(WA1-WA2)
4770 WA6=WA3-((BETA-BA1)/(BA2-BA1))*(WA3-WA4)
4780 WA=WA6-((UA2-UA)/(UA2-UA1))*(WA6-WA5)
4790 GOTO 5100
4800 WA1=WAT(FLAG1,8):WA2=WAT(FLAG2,8)
4810 WA=WA2-((UA2-UA)/(UA2-UA1))*(WA2-WA1)
4820 GOTO 5100
4830 WA1=WAT(19,FLAG3):WA2=WAT(19,FLAG4)
4840 WA=WA1-((BETA-BA1)/(BA2-BA1))*(WA1-WA2)
4850 GOTO 5100
4860 IF WUB <> 2 THEN 5100
4870 'IDENTIFYING CLOSEST 1/UB VALUES
4880 FOR J=1 TO 32
4890 FLAG1=J:FLAG2=J+1
4900 IF UB>=UBT(J) AND UB<=UBT(J+1) THEN 4920
4910 NEXT J
4920 UB1=UBT(FLAG1):UB2=UBT(FLAG2)
4930 'INTERPOLATION OF 1/UB AND W(1/UB,BETA) VALUES
4940 IF BETA=BETA1(8) AND UB=UBT(33) THEN WB= WBT(33,8)
4950 IF BETA=BETA1(8) AND UB=UBT(33) THEN 5100
4960 IF BETA=BETA1(8) THEN 5040
```

```
4970 IF UB=UBT(33) THEN 5070
4980 WB1=WBT(FLAG1,FLAG3):WB2=WBT(FLAG1,FLAG4)
4990 WB3=WBT(FLAG2,FLAG3):WB4=WBT(FLAG2,FLAG4)
5000 WB5=WB1-((BETA-BA1)/(BA2-BA1))*(WB1-WB2)
5010 WB6=WB3-((BETA-BA1)/(BA2-BA1))*(WB3-WB4)
5020 WB=WB6-((UB2-UB)/(UB2-UB1))*(WB6-WB5)
5030 GOTO 5100
5040 WB1=WBT(FLAG1,8):WB2=WBT(FLAG2,8)
5050 WB=WB2-((UB2-UB)/(UB2-UB1))*(WB2-WB1)
5060 GOTO 5100
5070 WB1=WBT(33,FLAG3):WB2=WBT(33,FLAG4)
5080 WB=WB1-((BETA-BA1)/(BA2-BA1))*(WB1-WB2)
5090 GOTO 5100
5100 RETURN
5110 'SUBROUTINE TO CALCULATE WFP(UF,R/MB,B,C)
5120 'USING APPROXIMATE EQUATIONS
5130 D=(2*UF)^.5/(RMB*B^.5)
5140 TD=((EXP(D)-EXP(-D))/2)/((EXP(D)+EXP(-D))/2)
5150 A=2*UF+RMB*C*(2*UF)^.5/B^.5*TD
5160 E=A^.5
5170 IF E<=2 THEN 5230
5180 L=2/E:M=L^2:N=L^3:O=L^4:P=L^5:S=L^6
5190 KO=1.25331414#-7.832358E-02*L+2.189568E-02*M-1.062446E-02*N
5200 KO=KO+5.87872E-03*O-.0025154*P+5.3208E-04*S
5210 KO=KO/(E^.5*EXP(E))
5220 GOTO 5290
5230 L=E/3.75:M=L^2:N=L^4:O=L^6:P=L^8:S=L^10:T=L^12
5240 IO=1+3.5156229#*M+3.0899424#*N+1.2067492#*O
5250 IO=IO+.2659732*P+.0360768*S+.0045813*T
5260 L=E/2:M=L^2:N=L^4:O=L^6:P=L^8:S=L^10:T=L^12
5270 KO=-LOG(L)*IO-.577721566#+.4227842#*M+.23069756#*N
5280 KO=KO+.0348859*O+.0026298*P+.0001075*S+.0000074*T
5290 WF=2*KO
5300 RETURN
5310 END
```

Appendix B
WELFLO Source Code

The source code for the WELFLO IBM PC BASICA microcomputer program is listed in this appendix. Program algorithms are documented as REM statements in the listing to assist the reader in understanding the code. Many statements in the program could be restructured or eliminated to produce more efficient coding, but they have been intentionally written in their present form to provide clear explanations of processing steps.

```
10 CLS:CLEAR:KEY OFF
20 AF$="######.##":AG$="#####.##":AT$="#####.###":AW$="###.##"
30 AP$="######.###":AD$="####.##":AA$="########.##":AS$="##.######"
40 DIM XWELL(50,25),YWELL(50,25),Q(50,25),TIME(50,25),XOB(30),YOB(30)
50 DIM DDO(25,25),IO(25),JO(25),DELTA(25),NWELLS(50),UFD(14)
60 DIM RAD(50,25),DD(30,30),WFD(14),DDB(25),DOB(25)
70 PRINT"Program: WELFLO"
80 PRINT"Version: IBM/PC BASICA 2.1"
90 PRINT"         Copyright 1988 Lewis Publishers, Inc."
100 PRINT"Purpose: Calculate drawdown or recovery values"
110 PRINT"Author : William C. Walton"
120 PRINT"Date   : April, 1988"
130 PRINT:PRINT"Do you have a printer?"
140 INPUT"Enter 1 for yes or 2 for no";AI
150 IF AI <> 1 AND AI <> 2 THEN 130
160 IF AI=2 THEN 200
170 PRINT:PRINT"Turn on printer or else an error will occur"
180 PRINT:PRINT"Press any key to continue"
190 P$=INKEY$:IF P$="" THEN 190
200 PRINT:PRINT"AQUIFER OPTIONS:"
210 PRINT:PRINT"It is assumed that aquitard storativity and"
220 PRINT"delayed gravity yield are negligible"
230 PRINT:PRINT"Enter 1 for nonleaky artesian"
240 PRINT"Enter 2 for leaky artesian"
250 PRINT"Enter 3 for water table"
260 INPUT"Enter 4 for nonleaky artesian fractured rock";AO
270 IF AO <> 1 AND AO <> 2 AND AO <>3 AND AO <> 4 THEN 230
280 PRINT:PRINT"WELL OPTIONS:"
290 PRINT:PRINT"It is assumed that well discharge rates are constant"
300 PRINT"during all time increments if wells partially penetrate the"
310 PRINT"aquifer and/or there is wellbore storage"
320 PRINT:PRINT"Enter 1 for fully penetrating wells"
330 PRINT"and no wellbore storage"
340 PRINT"Enter 2 for partially penetrating wells"
350 PRINT"Enter 3 for wellbore storage"
360 PRINT"Enter 4 for partially penetrating wells"
```

```
370 INPUT"and wellbore storage";WO
380 IF WO <> 1 AND WO <> 2 AND WO <> 3 AND WO <> 4 THEN 320
390 IF WO <> 2 AND WO <> 4 THEN 400
400 PRINT
410 IF AI=1 THEN LPRINT
420 PRINT"GENERAL DATA BASE:":PRINT
430 IF AI=1 THEN LPRINT"GENERAL DATA BASE:"
440 IF AI=1 THEN LPRINT
450 PRINT:PRINT"Durations of simulation periods should be selected"
460 PRINT"so that the times at the ends of periods are in ascending"
470 PRINT"order. The differences between these times should be small"
480 PRINT"if wellbore storage is simulated and high drawdown"
490 PRINT"calculation precision is required":PRINT
500 PRINT"Number of simulation periods for which drawdown"
510 INPUT"or recovery is to be calculated (must be <26)";TS
520 IF AI=2 THEN 550
530 LPRINT"Number of simulation periods for which drawdown"
540 LPRINT"or recovery is to be calculated";TS
550 PRINT:PRINT"Graphs of well discharge (+)or recharge (-) versus"
560 PRINT"time should be drawn for each production and injection"
570 PRINT"well and combined into simulation period pump operation"
580 PRINT"schedules. The duration of each simulation period is equal"
590 PRINT"to the longest pump operation period during each"
600 PRINT"simulation period":PRINT
610 FOR I=1 TO TS
620 PRINT"Simulation period number=";I
630 IF AI=1 THEN LPRINT"Simulation period number=";I
640 INPUT"Duration of simulation period in days=";DELTA(I)
650 IF AI=2 THEN 670
660 LPRINT"Duration of simulation period in days=" USING AT$;DELTA(I)
670 NEXT I
680 PRINT:PRINT"Any aquifer boundaries and/or discontinuities and associated"
690 PRINT"image wells and any injection wells should be drawn to"
700 PRINT"scale on a map. An area of interest which lies entirely"
710 PRINT"within aquifer boundaries and discontinuities should"
720 PRINT"be selected for drawdown or recovery calculations"
730 PRINT"and display. A uniform square grid should be superposed"
740 PRINT"over this area. Grid lines should be indexed using"
750 PRINT"the I (column), J (row) notation colinear with the"
760 PRINT"X and Y directions, respectively. I coordinates should"
770 PRINT"increase left to right and J coordinates should"
780 PRINT"increase top to bottom. The origin of the grid should"
```

```
790 PRINT"be beyond the grid in the upper-left corner of the map.":PRINT
800 PRINT:PRINT"Press any key to continue"
810 T$=INKEY$:IF T$="" THEN 810
820 PRINT
830 INPUT"Number of grid columns (must be <31)=";NC
840 INPUT"Number of grid rows(must be <31)=";NR
850 INPUT"grid spacing in ft=";GS
860 INPUT"X-coordinate of upper-left grid node in ft=";XGO
870 INPUT"Y-coordinate of upper-left grid node in ft=";YGO
880 FOR I=1 TO TS
890 PRINT:PRINT"Simulation period number=";I
900 PRINT"Number of production, injection, and image wells"
910 INPUT"active during simulation period (must be <51)=";NWELLS(I)
920 NEXT I
930 FOR J=1 TO TS
940 FOR I=1 TO NWELLS(J)
950 PRINT:PRINT"Well Number=";I
960 PRINT"Simulation period number=";J
970 INPUT"X-coordinate of well in ft=";XWELL(I,J)
980 INPUT"Y-coordinate of well in ft=";YWELL(I,J)
990 INPUT"Well discharge in gpm (enter negative number for recharge)=";Q(I,J)
1000 PRINT"Duration of pump operation during simulation"
1010 INPUT"period in days=";TIME(I,J)
1020 INPUT"Well radius in ft=";RAD(I,J)
1030 NEXT I,J
1040 IF AI=2 THEN 1230
1050 LPRINT"Number of grid columns=";NC
1060 LPRINT"Number of grid rows=";NR
1070 LPRINT"Grid spacing in ft=" USING AF$;GS
1080 LPRINT"X-coordinate of upper-left grid node in ft=" USING AF$;XGO
1090 LPRINT"Y-coordinate of upper-left grid node in ft=" USING AF$;YGO
1100 FOR J=1 TO TS
1110 FOR I=1 TO NWELLS(J)
1120 LPRINT"Simulation period number=";J
1130 LPRINT"Number of production, injection, and image wells"
1140 LPRINT"active during simulation period=";NWELLS(J)
1150 LPRINT"Well number=";I
1160 LPRINT"X-coordinate of well in ft=" USING AF$;XWELL(I,J)
1170 LPRINT"Y-coordinate of well in ft=" USING AF$;YWELL(I,J)
1180 LPRINT"Well discharge in gpm=" USING AG$;Q(I,J)
1190 LPRINT"Duration of pump operation during simulation period"
1200 LPRINT"in days=" USING AT$;TIME(I,J)
```

```
1210 LPRINT"Well radius in ft=" USING AW$;RAD(I,J)
1220 NEXT I,J
1230 PRINT:PRINT"Observation wells are located at grid nodes":PRINT
1240 PRINT"Number of observation wells for which time-"
1250 INPUT"drawdown tables are desired (must be <26)";NOBS
1260 IF AI=2 THEN 1290
1270 LPRINT"Number of observation wells for which time-"
1280 LPRINT"drawdown tables are desired";NOBS
1290 FOR I=1 TO NOBS
1300 PRINT
1310 PRINT"Observation well number=";I
1320 IF AI=1 THEN LPRINT"Observation well number=";I
1330 INPUT"I-coordinate of observation well=";IO(I)
1340 INPUT"J-coordinate of observation well=";JO(I)
1350 IF AI=2 THEN 1380
1360 LPRINT"I-coordinate of observation well=";IO(I)
1370 LPRINT"J-coordinate of observation well=";JO(I)
1380 NEXT I
1390 PRINT
1400 FOR J=1 TO TS
1410 FOR I=1 TO NOBS
1420 DDO(I,J)=0
1430 NEXT I,J
1440 IF AO=4 THEN 1640
1450 INPUT"Aquifer transmissivity in gpd/ft=";T
1460 IF AI=2 THEN 1480
1470 LPRINT"Aquifer transmissivity in gpd/ft=" USING AA$;T
1480 IF AO=3 THEN 1530
1490 INPUT"Aquifer storativity as a decimal=";STOR
1500 IF AI=2 THEN 1520
1510 LPRINT"Aquifer storativity as a decimal=" USING AS$;STOR
1520 GOTO 1560
1530 INPUT"Aquifer specific yield as a decimal=";STOR1
1540 IF AI=2 THEN 1560
1550 LPRINT"Aquifer specific yield as a decimal=" USING AS$;STOR1
1560 IF AO <> 2 THEN 1630
1570 INPUT"Aquitard thickness in ft=";MA
1580 IF AI=2 THEN 1600
1590 LPRINT"Aquitard thickness in ft=" USING AD$;MA
1600 INPUT"Aquitard vert. hydr. conduct. in gpd/sq ft=";PT
1610 IF AI=2 THEN 1630
1620 LPRINT"Aquitard vert. hydr. conduct. in gpd/sq ft=" USING AP$;PT
```

```
1630 IF AO <> 4 THEN 1880
1640 PRINT"Fissure horizontal hydraulic"
1650 INPUT"conductivity in gpd/sq ft=";TFP
1660 INPUT"Fractured rock aquifer thickness in ft=";MRT
1670 TF=TFP*MRT
1680 PRINT"Block vertical hydraulic"
1690 INPUT"conductivity in gpd/sq ft=";TB
1700 PRINT"Storativity of fissured portion of"
1710 INPUT"fractured rock aquifer as a decimal=";SF
1720 PRINT"Storativity of block portion of"
1730 INPUT"fractured rock aquifer as a decimal=";SB
1740 INPUT"Half dimension (thickness) of average block unit in ft=";MB
1750 IF AI=2 THEN 1860
1760 LPRINT"Fissure horizontal hydraulic"
1770 LPRINT"conductivity in gpd/sq ft=" USING AP$;TFP
1780 LPRINT"Fractured rock aquifer thickness in ft=" USING AD$;MRT
1790 LPRINT"Block vertical hydraulic"
1800 LPRINT"conductivity in gpd/sq ft=" USING AS$;TB
1810 LPRINT"Storativity of fissured portion of"
1820 LPRINT"fractured rock aquifer as a decimal=" USING AS$;SF
1830 LPRINT"Storativity of block portion of"
1840 LPRINT"fractured rock aquifer as a decimal=" USING AS$;SB
1850 LPRINT"Half dimension of average block unit in ft=" USING AD$;MB
1860 B=TB*SF/(SB*TFP)
1870 C=TB/TFP
1880 PRINT:PRINT"Enter Y to revise general data base"
1890 INPUT"or N to continue";R$:PRINT
1900 IF R$<>"N" AND R$<>"n" AND R$<>"Y" AND R$<>"y" THEN 1880
1910 IF R$="Y" OR R$="y" THEN 200
1920 PRINT
1930 IF AI=1 THEN LPRINT
1940 'COMPUTE COORDINATES OF GRID NODES
1950 FOR J=1 TO NR
1960 CLS:PRINT"COMPUTATIONS ARE IN PROGRESS"
1970 FOR I=1 TO NC
1980 IF I=1 THEN 2010
1990 XOB(I)=GS*(I-1)+XGO
2000 GOTO 2020
2010 XOB(I)=XGO
2020 IF J=1 THEN 2050
2030 YOB(J)=GS*(J-1)+YGO
2040 GOTO 2060
```

```
2050 YOB(J)=YGO
2060 NEXT I,J
2070 'CALCULATE DISTANCES BETWEEN WELLS AND GRID NODES,
2080 'WELL FUNCTIONS, AND DRAWDOWNS
2090 FOR M=1 TO TS
2100 FOR K=1 TO NR
2110 FOR J=1 TO NC
2120 DD(J,K)=0!
2130 FOR I=1 TO NWELLS(M)
2140 CLS:PRINT"COMPUTATIONS ARE IN PROGRESS"
2150 R2=ABS(XOB(J)-XWELL(I,M))^2+ABS(YOB(K)-YWELL(I,M))^2
2160 IF R2=0 THEN R2=RAD(I,M)^2
2170 IF AO <> 1 AND AO <> 2 THEN 2190
2180 U=1.87*R2*STOR/(T*TIME(I,M))
2190 IF AO <> 2 THEN 2210
2200 S=R2^.5/((T/(PT/MA))^.5)
2210 IF AO <> 3 THEN 2230
2220 U=1.87*R2*STOR1/(T*TIME(I,M))
2230 IF AO <> 4 THEN 2260
2240 UF=1.87*R2*SF/(TF*TIME(I,M))
2250 RMB=R2^.5/MB
2260 IF AO <> 1 THEN 2290
2270 GOSUB 4210
2280 DD(J,K)=DD(J,K)+114.6*Q(I,M)*W/T
2290 IF AO <> 2 THEN 2320
2300 GOSUB 4320
2310 DD(J,K)=DD(J,K)+114.6*Q(I,M)*WS/T
2320 IF AO <> 3 THEN 2350
2330 GOSUB 4210
2340 DD(J,K)=DD(J,K)+114.6*Q(I,M)*W/T
2350 IF AO <> 4 THEN 2380
2360 GOSUB 4840
2370 DD(J,K)=DD(J,K)+114.6*Q(I,M)*WF/TF
2380 NEXT I,J,K
2390 FOR I=1 TO NOBS
2400 FOR K=1 TO NR
2410 FOR J=1 TO NC
2420 IF J=IO(I) AND K=JO(I) THEN DDO(I,M)=DD(J,K)
2430 NEXT J,K,I
2440 PRINT:PRINT"Nodal computation results will now be"
2450 PRINT"displayed in tabular form, use Ctrl-Num Lock"
2460 PRINT"keys to control screen scrolling"
```

```
2470 PRINT
2480 IF AI=1 THEN LPRINT
2490 PRINT"Press any key to continue"
2500 T$=INKEY$:IF T$="" THEN 2500
2510 CLS:PRINT:PRINT"NODAL COMPUTATION RESULTS:"
2520 IF AI=1 THEN LPRINT"NODAL COMPUTATION RESULTS:"
2530 PRINT
2540 IF AI=1 THEN LPRINT
2550 PRINT"SIMULATION PERIOD DURATION IN DAYS:" USING AT$;DELTA(M)
2560 IF AI=2 THEN 2580
2570 LPRINT"SIMULATION PERIOD DURATION IN DAYS:" USING AT$;DELTA(M)
2580 PRINT
2590 IF AI=1 THEN LPRINT
2600 PRINT"VALUES OF DRAWDOWN OR RECOVERY (FT) AT NODES:"
2610 IF AI=2 THEN 2630
2620 LPRINT"VALUES OF DRAWDOWN OR RECOVERY (FT) AT NODES:"
2630 PRINT
2640 IF AI=1 THEN LPRINT
2650 PRINT"J-ROW " SPC(26) "I-COLUMN"
2660 PRINT"        1      2      3      4      5";
2670 PRINT"      6      7      8      9      10":PRINT
2680 IF AI=2 THEN 2720
2690 LPRINT"J-ROW " SPC(26) "I-COLUMN"
2700 LPRINT"        1      2      3      4      5";
2710 LPRINT"      6      7      8      9      10":LPRINT
2720 IF NC=10 OR NC>10 THEN NCC=10
2730 IF NC<10 THEN NCC=NC
2740 FOR J=1 TO NR
2750 IF J>9 THEN 2780
2760 PRINT" ";J;
2770 GOTO 2790
2780 PRINT"";J;
2790 FOR I=1 TO NCC
2800 IF I=NCC THEN PRINT USING AD$;DD(I,J)
2810 IF I=NCC THEN 2830
2820 PRINT USING AD$;DD(I,J);
2830 NEXT I,J:PRINT
2840 IF NC<10 THEN NCC=NC
2850 IF NC=10 OR NC>10 THEN NCC=10
2860 FOR J=1 TO NR
2870 IF J>9 THEN 2900
2880 IF AI=1 THEN LPRINT" ";J;
```

```
2890 GOTO 2910
2900 IF AI=1 THEN LPRINT"";J;
2910 FOR I=1 TO NCC
2920 IF AI=2 THEN 2940
2930 IF I=NCC THEN LPRINT USING AD$;DD(I,J)
2940 IF I=NCC THEN 2970
2950 IF AI=2 THEN 2970
2960 IF AI=1 THEN LPRINT USING AD$;DD(I,J);
2970 NEXT I,J:PRINT
2980 IF AI=1 THEN LPRINT
2990 IF NC<11 THEN 3510
3000 PRINT"J-ROW " SPC(26) "I-COLUMN"
3010 PRINT"        11     12     13     14     15";
3020 PRINT"     16     17     18     19     20":PRINT
3030 IF AI=2 THEN 3070
3040 LPRINT"J-ROW " SPC(26) "I-COLUMN"
3050 LPRINT"        11     12     13     14     15";
3060 LPRINT"     16     17     18     19     20":LPRINT
3070 IF NC<20 THEN NCC=NC
3080 IF NC=20 OR NC>20 THEN NCC=20
3090 FOR J=1 TO NR
3100 IF J>9 THEN 3140
3110 PRINT" ";J;
3120 IF AI=1 THEN LPRINT" ";J;
3130 GOTO 3160
3140 PRINT"";J;
3150 IF AI=1 THEN LPRINT"";J;
3160 FOR I=11 TO NCC
3170 IF I=NCC THEN PRINT USING AD$;DD(I,J)
3180 IF AI=2 THEN 3200
3190 IF I=NCC THEN LPRINT USING AD$;DD(I,J)
3200 IF I=NCC THEN 3230
3210 PRINT USING AD$;DD(I,J);
3220 IF AI=1 THEN LPRINT USING AD$;DD(I,J);
3230 NEXT I,J:PRINT
3240 IF AI=1 THEN LPRINT
3250 IF NC<21 THEN 3510
3260 PRINT"J-ROW " SPC(26) "I-COLUMN"
3270 PRINT"        21     22     23     24     25";
3280 PRINT"     26     27     28     29     30":PRINT
3290 IF AI=2 THEN 3330
3300 LPRINT"J-ROW " SPC(26) "I-COLUMN"
```

```
3310 LPRINT"      21     22     23     24     25";
3320 LPRINT"      26     27     28     29     30":LPRINT
3330 IF NC<30 THEN NCC=NC
3340 IF NC=30 THEN NCC=30
3350 FOR J=1 TO NR
3360 IF J>9 THEN 3400
3370 PRINT" ";J;
3380 IF AI=1 THEN LPRINT" ";J;
3390 GOTO 3420
3400 PRINT"";J;
3410 IF AI=1 THEN LPRINT"";J;
3420 FOR I=21 TO NCC
3430 IF I=NCC THEN PRINT USING AD$;DD(I,J)
3440 IF AI=2 THEN 3460
3450 IF I=NCC THEN LPRINT USING AD$;DD(I,J)
3460 IF I=NCC THEN 3490
3470 PRINT USING AD$;DD(I,J);
3480 IF AI=1 THEN LPRINT USING AD$;DD(I,J);
3490 NEXT I,J
3500 IF AI=1 THEN LPRINT
3510 PRINT
3520 PRINT"Enter Y to create a sequential data file of results"
3530 PRINT"for export to graphics programs GWGRAF or OMNIPLOT."
3540 INPUT"Enter N for no file";CD$:PRINT
3550 IF CD$<>"N" AND CD$<>"n" AND CD$<>"Y" AND CD$<>"y" THEN 3520
3560 IF CD$="N" OR CD$="n" THEN 3640
3570 INPUT"Enter a valid filename with extension .DAT such as FILE1.DAT";F$
3580 OPEN F$ FOR OUTPUT AS #1
3590 FOR J=1 TO NR
3600 FOR I=1 TO NC
3610 PRINT #1, USING"####.##";-DD(I,J)
3620 NEXT I,J
3630 CLOSE #1
3640 PRINT
3650 PRINT"Enter Y to create a sequential data file of results"
3660 PRINT"for export to graphics program SURFER."
3670 INPUT"Enter N for no file";CD$:PRINT
3680 IF CD$<>"N" AND CD$<>"n" AND CD$<>"Y" AND CD$<>"y" THEN 3650
3690 IF CD$="N" OR CD$="n" THEN 3770
3700 INPUT"Enter a valid filename with extension .DAT such as FILE1.DAT";F$
3710 OPEN F$ FOR OUTPUT AS #1
3720 FOR J=1 TO NR
```

```
3730 FOR I=1 TO NC
3740 WRITE #1,XOB(I),YOB(J),-DD(I,J)
3750 NEXT I,J
3760 CLOSE #1
3770 IF WO <> 2 AND WO <> 4 THEN 3790
3780 GOSUB 5020
3790 IF WO <> 3 AND WO <> 4 THEN 3810
3800 GOSUB 5820
3810 PRINT
3820 NEXT M
3830 FOR J=1 TO NOBS
3840 IF J > 1 THEN 4030
3850 PRINT:PRINT"Obs. well computation results will now be"
3860 PRINT"displayed in tabular form, use Ctrl-Num Lock"
3870 PRINT"keys to control screen scrolling"
3880 PRINT
3890 IF AI=1 THEN LPRINT
3900 PRINT"Press any key to continue"
3910 T$=INKEY$:IF T$="" THEN 3910
3920 CLS:PRINT
3930 IF AI=1 THEN LPRINT
3940 PRINT"OBSERVATION WELL COMPUTATION RESULTS WITH FULL WELL"
3950 PRINT"PENETRATION AND WITHOUT WELLBORE STORAGE:":PRINT
3960 IF AI=2 THEN 3990
3970 LPRINT"OBSERVATION WELL COMPUTATION RESULTS WITH FULL WELL"
3980 LPRINT"PENETRATION AND WITHOUT WELLBORE STORAGE:":LPRINT
3990 PRINT
4000 IF AI=1 THEN LPRINT
4010 PRINT"TIME-DRAWDOWN OR RECOVERY TABLE"
4020 IF AI=1 THEN LPRINT"TIME-DRAWDOWN OR RECOVERY TABLE"
4030 PRINT
4040 IF AI=1 THEN LPRINT
4050 PRINT"OBSERVATION WELL NUMBER:";J
4060 IF AI=1 THEN LPRINT"OBSERVATION WELL NUMBER:";J
4070 PRINT
4080 IF AI=1 THEN LPRINT
4090 PRINT"TIME(DAYS)        DRAWDOWN OR RECOVERY(FT)"
4100 IF AI=1 THEN LPRINT"TIME(DAYS)        DRAWDOWN OR RECOVERY(FT)"
4110 FOR I=1 TO TS
4120 PRINT USING"#####.###";DELTA(I);
4130 PRINT"
4140 PRINT USING"#########.##";DDO(J,I)
```

```
4150 IF AI=1 THEN LPRINT USING"#####.###";DELTA(I);
4160 IF AI=1 THEN LPRINT"        ";
4170 IF AI=1 THEN LPRINT USING"#########.##";DDO(J,I)
4180 NEXT I,J
4190 END
4200 'SUBROUTINE TO COMPUTE W(U) USING POLYNOMIAL APPROXIMATIONS
4210 A1=U^2:B1=U^3:C1=U^4:D1=U^5
4220 IF U > 10 THEN W=0
4230 IF U > 10 THEN 4260
4240 IF U>1 THEN GOTO 4270
4250 W=-LOG(U)-.5772156600000006#+.9999919300000007#*U-.24991055#*A1+.0551997*B1
-9.76004E-03*C1+1.07857E-03*D1
4260 RETURN
4270 W=C1+8.573328740100012#*B1+18.059016973#*A1+8.634760892499999#*U+.267773734
3000006#
4280 W=W/(C1+9.573322345400008#*B1+25.6329561486#*A1+21.0996530827#*U+3.95849692
2800006#)
4290 W=W/(U*EXP(U))
4300 RETURN
4310 'SUBROUTINE TO COMPUTE W(U,R/B) USING POLYNOMIAL APPROXIMATIONS
4320 IF U>10! THEN WS=0:GOTO 4790
4330 IF S>2 THEN 4690
4340 IF U>=1 THEN 4370
4350 WU=-LOG(U)-.5772156600000009#+.9999919300000017#*U-.24991055#*U*U
4360 WU=WU+.0551997*U^3-9.76004E-03*U^4+1.07857E-03*U^5:GOTO 4420
4370 WU=U^4+8.573328740100025#*U^3+18.059016973#*U^2+8.634760892499999#*U
4380 WU=WU+.2677737343#
4390 WUP=U^4+9.573322345400009#*U^3+25.6329561486#*U^2+21.0996530827#*U
4400 WUP=WUP+3.95849692280001#
4410 WU=WU/WUP
4420 L=S/3.75
4430 V=1+3.5156229#*L^2+3.0899424#*L^4+1.2067492#*L^6+.265973*L^8+.0360768*L^10
4440 V=V+.0045813*L^12
4450 F=S/2
4460 H=-LOG(F)*V-.5772156600000009#+.422784*F^2+.23069756#*F^4+.0348859*F^6
4470 H=H+2.62698E-03*F^8+.0001075*F^10+.0000074*F^12
4480 IF S=0 THEN WS=WU:GOTO 4790
4490 N=S^2/(4*U)
4500 IF N>5 THEN WS=2*H:GOTO 4790
4510 IF U<=.9 THEN 4540
4520 A=U+.5858:B=U+3.414:C=S*S/4
4530 WS=1.5637*EXP(-A-C/A)/A+4.54*EXP(-B-C/B)/B:GOTO 4790
```

```
4540 IF U<.05 THEN 4580
4550 IF U>S/2 THEN 4570
4560 C=-(1.75*U)^-.448*S:WS=2*H-4.8*10^C:GOTO 4790
4570 WS=WU-(S/(4.7*U^.6))^2:GOTO 4790
4580 IF U>.01 AND S<.1 THEN 4570
4590 IF N<1 THEN 4660
4600 WN=N^4+8.573328740100025#*N^3+18.059016973#*N^2+8.634760892499999#*N
4610 WN=WN+.267773734300001#
4620 WNP=N^4+9.573322345400009#*N^3+25.6329561486#*N^2+21.0996530827#*N
4630 WNP=WNP+3.958496922800012#
4640 WN=WN/WNP
4650 WN=WN/(N*EXP(N)):GOTO 4680
4660 WN=-LOG(N)+.9999919300000017#*N-.5772156600000009#-.24991055#*N^2
4670 WN=WN+.0551997*N^3-9.76004E-03*N^4+1.07857E-03*N^5
4680 WS=2*H-WN*V:GOTO 4790
4690 MX=-((S-2*U)/(2*U^.5))
4700 IF MX<0 THEN MX=ABS(MX):GOSUB 4720:GOTO 4780
4710 GOSUB 4720:GOTO 4770
4720 NP=1+.0705230784#*MX+.0422820123#*MX^2+9.270527200000026D-03*MX^3
4730 NP=NP+1.52014E-04*MX^4+2.76567E-04*MX^5+4.30638E-05*MX^6
4740 IF NP>100! THEN 4760
4750 N=1/NP^16:RETURN
4760 N=0!:RETURN
4770 WS=(3.1416/(2*S))^.5*EXP(-S)*N:GOTO 4790
4780 WS=(3.1416/(2*S))^.5*EXP(-S)*(2-N)
4790 RETURN
4800 'SUBROUTINE TO CALCULATE WFP(UF,R/MB,B,C)
4810 'USING APPROXIMATE EQUATIONS
4820 IF UF > 10 THEN WF=0
4830 IF UF > 10 THEN 5010
4840 D=(2*UF)^.5/(RMB*B^.5)
4850 TD=((EXP(D)-EXP(-D))/2)/((EXP(D)+EXP(-D))/2)
4860 A=2*UF+RMB*C*(2*UF)^.5/B^.5*TD
4870 E=A^.5
4880 IF E<=2 THEN 4940
4890 L1=2/E:M1=L1^2:N1=L1^3:O1=L1^4:P1=L1^5:S1=L1^6
4900 KO=1.25331414#-7.832358E-02*L1+2.189568E-02*M1-1.062446E-02*N1
4910 KO=KO+5.87872E-03*O1-.0025154*P1+5.3208E-04*S1
4920 KO=KO/(E^.5*EXP(E))
4930 GOTO 5000
4940 L2=E/3.75:M2=L2^2:N2=L2^4:O2=L2^6:P2=L2^8:S2=L2^10:T2=L2^12
4950 IO=1+3.5156229#*M2+3.0899424#*N2+1.2067492#*O2
```

```
4960 IO=IO+.2659732*P2+.0360768*S2+.0045813*T2
4970 L3=E/2:M3=L3^2:N3=L3^4:O3=L3^6:P3=L3^8:S3=L3^10:T3=L3^12
4980 KO=-LOG(L3)*IO-.577721566#+.4227842#*M3+.23069756#*N3
4990 KO=KO+.0348859*O3+.0026298*P3+.0001075*S3+.0000074*T3
5000 WF=2*KO
5010 RETURN
5020 PRINT
5030 PRINT"Enter 1 to calculate partial penetration"
5040 PRINT"impacts in production well"
5050 PRINT"Enter 2 to calculate partial penetration"
5060 INPUT"impacts in observation well";PPW
5070 IF PPW <> 1 AND PPW <> 2 THEN 5030
5080 PRINT
5090 IF AI=1 THEN LPRINT
5100 PRINT"WELL PARTIAL PENETRATION DATA BASE:"
5110 IF AI=1 THEN LPRINT"WELL PARTIAL PENETRATION DATA BASE:"
5120 PRINT
5130 IF AI=1 THEN LPRINT
5140 PRINT"Drawdown with full well penetration as"
5150 INPUT"calculated earlier in this program in ft=";DWF
5160 IF AI=2 THEN 5190
5170 LPRINT"Drawdown with full well penetration as"
5180 LPRINT"calculated earlier in this program in ft=";DWF
5190 INPUT"Production well discharge rate in gpm=";QWF
5200 IF AI=2 THEN 5220
5210 LPRINT"Production well discharge rate in gpm=";QWF
5220 IF PPW=1 THEN 5260
5230 INPUT"Radial distance to observation well in ft=";RD
5240 IF AI=2 THEN 5260
5250 LPRINT"Radial distance to observation well in ft=" USING AF$;RD
5260 INPUT"Aquifer thickness in ft=";THICK
5270 IF AI=2 THEN 5290
5280 LPRINT"Aquifer thickness in ft=" USING AD$;THICK
5290 INPUT"Aquifer horiz. hydraulic conductivity in gpd/sq ft=";PH
5300 IF AI=2 THEN 5320
5310 LPRINT"Aquifer horiz. hydr. conduct. in gpd/sq ft=" USING AP$;PH
5320 INPUT"Aquifer vert. hydraulic conductivity in gpd/sq ft=";PV
5330 IF AI=2 THEN 5350
5340 LPRINT"Aquifer vert. hydr. conduct. in gpd/sq ft=" USING AP$;PV
5350 INPUT"Dist. from aquifer top to prod. well bottom in ft=";LP
5360 IF AI=2 THEN 5380
5370 LPRINT"Dist. from aquifer top to prod. well bottom(ft)" USING AD$;LP
```

```
5380 INPUT"Dist. from aquifer top to prod. well screen top in ft=";DP
5390 IF AI=2 THEN 5410
5400 LPRINT"Dist. from aquifer top to prod. well screen top (ft)=" USING AD$;DP
5410 IF PPW=1 THEN 5490
5420 INPUT"Dist. from aquifer top to obs. well bottom in ft="; LL
5430 IF AI=2 THEN 5450
5440 LPRINT"Dist. from aquifer top to obs. well bottom in ft=" USING AD$;LL
5450 INPUT"Dist. from aquifer top to obs. well screen top in ft=";DD
5460 IF AI=2 THEN 5480
5470 LPRINT"Dist. from aquifer top to obs. well screen top (ft)=" USING AD$;DD
5480 GOTO 5550
5490 INPUT"Production well radius in ft=";RD
5500 IF AI=2 THEN 5520
5510 LPRINT"Production well radius in ft=" USING AD$;RD
5520 LL=(LP+DP)/2:DD=LP-1
5530 'CALCULATE WELL FUNCTION USING WELL PARTIAL
5540 'PENETRATION IMPACT APPROXIMATION
5550 Z=0
5560 IF AO=1 OR AO=2 THEN U=1.87*RD^2*STOR/(T*DELTA(M))
5570 IF AO=3 THEN U=1.87*RD^2*STOR1/(T*DELTA(M))
5580 IF AO=4 THEN U=1.87*RD^2*SF/(TF*DELTA(M))
5590 FOR K= 1 TO 50
5600 CLS:PRINT:PRINT"COMPUTATIONS ARE IN PROGRESS"
5610 S=K*3.1416*RD*(PV/PH)^.5/THICK
5620 GOSUB 4320
5630 AA=SIN(K*3.1416*(LP/THICK))-SIN(K*3.1416*(DP/THICK))
5640 BB=SIN(K*3.1416*(LL/THICK))-SIN(K*3.1416*(DD/THICK))
5650 Z=Z+1/K^2*(AA*BB*WS)
5660 NEXT K
5670 WUPAR=2*THICK^2*Z/(3.1416^2*(LP-DP)*(LL-DD))
5680 IF AO=4 THEN TRANS=TF
5690 IF AO <> 4 THEN TRANS=T
5700 DWP=DWF+114.6*QWF*WUPAR/TRANS
5710 PRINT:PRINT"PARTIAL PENETRATION COMPUTATION RESULTS:":PRINT
5720 IF AI=2 THEN 5740
5730 LPRINT:LPRINT"PARTIAL PENETRATION COMPUTATION RESULTS:":LPRINT
5740 PRINT"Drawdown with partially penetration in ft=" USING AD$;DWP
5750 IF AI=2 THEN 5770
5760 LPRINT"Drawdown with partially penetration in ft=" USING AD$;DWP
5770 PRINT:PRINT"Enter 1 for another well partial penetration"
5780 INPUT"calculation. Enter 2 to continue program";PPP
5790 IF PPP <> 1 AND PPP <> 2 THEN 5770
```

```
5800 IF PPP=1 THEN 5020
5810 RETURN
5820 PRINT
5830 IF AI=1 THEN LPRINT
5840 PRINT"WELLBORE STORAGE DATA BASE:"
5850 IF AI=1 THEN LPRINT"WELLBORE STORAGE DATA BASE:"
5860 PRINT
5870 IF AI=1 THEN LPRINT
5880 PRINT"Number of wells for which wellbore storage"
5890 INPUT"impacts are to be calculated (must be < 26)=";NW
5900 IF M>1 THEN 5940
5910 FOR K=1 TO NW
5920 DDB(K)=0:DOB(K)=0
5930 NEXT K
5940 FOR K=1 TO NW
5950 PRINT"Enter 1 if well in question is a production well"
5960 INPUT"Enter 2 if well in question is an observation well";WWB
5970 IF WWB <> 1 AND WWB <> 2 THEN 5950
5980 PRINT
5990 IF M > 1 THEN 6080
6000 PRINT"Production well discharge rate during the"
6010 INPUT"present simulation period in gpm=";QWB
6020 IF AI=2 THEN 6050
6030 LPRINT"Production well discharge rate during the"
6040 LPRINT"present simulation period in gpm=" USING AG$;QWB
6050 INPUT"Production well radius in ft=";RWB
6060 IF AI=2 THEN 6080
6070 LPRINT"Production well radius in ft=" USING AW$;RWB
6080 IF M=1 THEN QAT1=0
6090 IF M=1 THEN 6110
6100 QAT1=QAT
6110 IF M > 1 THEN 6150
6120 INPUT"Pump-column pipe radius in ft=";RCB
6130 IF AI=2 THEN 6150
6140 LPRINT"Pump-column pipe radius in ft=" USING AW$;RCB
6150 PRINT"Production well drawdown with full or partial penetration"
6160 PRINT"and without wellbore storage as calculated earlier"
6170 PRINT"in this program at the end of present simulation"
6180 INPUT"period in ft=";DDWB
6190 IF M=1 THEN DDWB1=0
6200 IF M=1 THEN 6220
6210 DDWB1=DDB(K)
```

```
6220 IF AI=2 THEN 6310
6230 LPRINT"Production well drawdown with full or partial penetration"
6240 LPRINT"and without wellbore storage as calculated earlier"
6250 LPRINT"in this program at the end of the present simulation"
6260 LPRINT"period in ft=" USING AD$;DDWB
6270 LPRINT"Production well drawdown with full or partial penetration"
6280 LPRINT"and without wellbore storage as calculated earlier"
6290 LPRINT"in this program at the end of the previous simulation"
6300 LPRINT"period in ft=" USING AD$;DDWB1
6310 IF WWB=1 THEN  6480
6320 PRINT"Drawdown in observation well with full or partial"
6330 PRINT"penetration and without wellbore storage as"
6340 PRINT"calculated earlier in this program at the end"
6350 INPUT"of the present simulation period in ft=";DDDB
6360 IF M=1 THEN DDDB1=0
6370 IF M=1 THEN 6390
6380 DDDB1=DOB(K)
6390 IF AI=2 THEN 6480
6400 LPRINT"Drawdown in observation well with full or partial"
6410 LPRINT"penetration and without wellbore storage as"
6420 LPRINT"calculated earlier in this program at the end"
6430 LPRINT"of the present simulation period in ft=" USING AD$;DDDB
6440 LPRINT"Drawdown in observation well with full or partial"
6450 LPRINT"penetration and without wellbore storage as"
6460 LPRINT"calculated earlier in this program at the end"
6470 LPRINT"of the previous simulation period in ft=" USING AD$;DDDB1
6480 IF M=1 THEN TPS1=0
6490 IF M=1 THEN 6510
6500 TPS1=DELTA(M-1)*1440
6510 TPS=DELTA(M)*1440
6520 IF AO=4 THEN TRANS=TF
6530 IF AO <> 4 THEN TRANS=T
6540 TZ=500000!*(RWB^2-RCB^2)/TRANS
6550 IF TPS > TZ THEN QAT=2*QWB
6560 IF TPS > TZ THEN 7220
6570 'CALCULATE DRAWDOWN IN FINITE DIAMETER PRODUCTION
6580 'WELL INSTEAD OF DRAWDOWN IN INFINITESIMAL DIAMETER
6590 'PRODUCTION WELL USING TABLED VALUES OF WELL FUNCTION
6600 FLAGT=1
6610 IF M=1 THEN 7060
6620 TWB1=TPS1
6630 IF AO <> 1 THEN 6670
```

```
6640 U=2693*RWB^2*STOR/(T*TWB1)
6650 GOSUB 4210
6660 SWB=114.6*QWB*W/T
6670 IF AO <> 2 THEN 6730
6680 U=2693*RWB^2*STOR/(T*TWB1)
6690 S=RWB/((T/(PT/MA))^.5)
6700 CLS:PRINT:PRINT"COMPUTATIONS ARE IN PROGRESS"
6710 GOSUB 4320
6720 SWB=114.6*QWB*WS/T
6730 IF AO <> 3 THEN 6770
6740 U=2693*RWB^2*STOR1/(T*TWB1)
6750 GOSUB 4210
6760 SWB=114.6*QWB*W/T
6770 IF AO <> 4 THEN 6820
6780 UF=2693*RWB^2*SF/(TF*TWB1)
6790 B=TB*SF/(TF*SB):C=TB/TF
6800 GOSUB 4840
6810 SWB=114.6*QWB*WF/TF
6820 U=1/U
6830 IF U>100 THEN 7120
6840 FOR I=1 TO 14
6850 READ UFD(I)
6860 NEXT I
6870 'VALUES OF 1/U IN TABLE
6880 DATA .1,.2,.4,.6,.8,1.0,2.0,4.0,6.0,8.0,10.0,20.0,50.0,100.0
6890 FOR I=1 TO 14
6900 READ WFD(I)
6910 NEXT I
6920 'VALUES OF W(U,R/RW) IN TABLE
6930 DATA 0.3353,0.4603,0.6308,0.7500,0.8496,0.9346,1.2328,1.6000
6940 DATA 1.8451,2.0366,2.1881,2.7249,3.4966,4.1241
6950 IF U<.1 THEN WD=0
6960 IF U<.1 THEN 7090
6970 FOR I=1 TO 13
6980 FLAG1=I:FLAG2=I+1
6990 IF U=UFD(14) THEN WD=WFD(14)
7000 IF U=UFD(14) THEN 7090
7010 WA1=WFD(FLAG1):WA2=WFD(FLAG2)
7020 WD=WA2-((UFD(FLAG2)-U)/(UFD(FLAG2)-UFD(FLAG1))*(WA2-WA1))
7030 NEXT I
7040 IF FLAGT>1 THEN 7090
7050 DDWB1=DDWB1-SWB+114.6*QWB*WD/TRANS
```

```
7060 TWB1=TPS
7070 FLAGT=FLAGT+1
7080 GOTO 6630
7090 DDWB=DDWB-SWB+114.6*QWB*WD/TRANS
7100 'CALCULATE WELLBORE STORAGE IMPACT USING METHOD
7110 'OF SUCCESSIVE APPROXIMATIONS OF DISCHARGE FROM AQUIFER
7120 S1=DDWB-DDWB1
7130 D1=0:D2=S1
7140 CS=(D1+D2)/2
7150 QAT=2*QWB-(2*3.1416*(RWB^2-RCB^2)*CS*7.48/(TPS-TPS1))-QAT1
7160 D3=(QAT+QAT1)*S1/(2*QWB)
7170 IF ABS(CS-D3)<.001 OR D1=D2 THEN 7220
7180 IF D3<CS THEN D2=CS
7190 IF D3>CS THEN D1=CS
7200 GOTO 7140
7210 IF K=1 THEN 7250
7220 IF WWB=1 THEN DDB(K)=DDB(K)+(DDWB-DDWB1)*(QAT+QAT1)/(2*QWB)
7230 IF WWB=2 THEN DOB(K)=DOB(K)+(DDDB-DDDB1)*(QAT+QAT1)/(2*QWB)
7240 GOTO 7270
7250 IF WWB=1 THEN DDB(K)=(DDWB-DDWB1)*(QAT+QAT1)/(2*QWB)
7260 IF WWB=2 THEN DDB(K)=(DDDB-DDDB1)*(QAT+QAT1)/(2*QWB)
7270 PRINT:PRINT"WELLBORE STORAGE COMPUTATION RESULTS:":PRINT
7280 IF AI=1 THEN LPRINT
7290 IF AI=1 THEN LPRINT"WELLBORE STORAGE COMPUTATION RESULTS:"
7300 IF AI=1 THEN LPRINT
7310 IF WWB=2 THEN 7370
7320 PRINT:PRINT"Drawdown with wellbore storage in ft=" USING AD$;DDB(K)
7330 IF AI=2 THEN 7360
7340 LPRINT
7350 LPRINT"Drawdown with wellbore storage in ft=" USING AD$;DDB(K)
7360 GOTO 7410
7370 PRINT:PRINT"Drawdown with wellbore storage in ft=" USING AD$;DOB(K)
7380 IF AI=2 THEN 7410
7390 LPRINT
7400 LPRINT"Drawdown with wellbore storage in ft=" USING AD$;DOB(K)
7410 PRINT
7420 NEXT K
7430 PRINT
7440 PRINT"Press any key to continue"
7450 B$=INKEY$:IF B$="" THEN 7450
7460 PRINT
7470 RETURN
```

Appendix C
CONMIG Source Code

The source code for the CONMIG IBM PC BASICA microcomputer program is listed in this appendix. Program algorithms are documented as REM statements in the listing to assist the reader in understanding the code. Many statements in the program could be restructured or eliminated to produce more efficient coding, but they have been intentionally written in their present form to provide clear explanations of processing steps.

```
10 CLS:CLEAR:KEY OFF
20 AT$="#####.##":AD$="######.##":AV$="####.##":AL$="####.###"
30 AY$="##.###":AK$="###.##":AC$="#####.###":A$="#####.##"
40 DIM DELTA(25),CS(30,25),CL(30,25),X(30,25),Y(30,25),Q(30,25),C(30,25)
50 DIM CM(30,25),VS(30,25),TIME(30,25),CCO(25,25),CC(30,30),XOB(30)
60 DIM IO(25),JO(25),DISPL(25),DISPT(25),VEL(25),NPTS(25),YOB(30)
70 DIM Z2(25)
80 PRINT"Program: CONMIG"
90 PRINT"Version: IBM/PC BASICA 2.1"
100 PRINT"         Copyright 1988 Lewis Publishers, Inc."
110 PRINT"Purpose: Calculation of contaminant"
120 PRINT"         concentration distribution"
130 PRINT"Author : William C. Walton"
140 PRINT"Date   : April, 1988"
150 PRINT:PRINT"Do you have a printer?"
160 INPUT"Enter 1 for yes or 2 for no";AI
170 IF AI <> 1 AND AI <> 2 THEN 150
180 IF AI=2 THEN 220
190 PRINT:PRINT"Turn on the printer or else an error will occur"
200 PRINT:PRINT"Press any key to continue"
210 P$=INKEY$:IF P$="" THEN 210
220 PRINT
230 PRINT"DATA BASE:":PRINT
240 IF AI=1 THEN LPRINT"DATA BASE:"
250 IF AI=1 THEN LPRINT
260 PRINT"Number of simulation periods for which contaminant"
270 PRINT"concentration distribution is to be calculated"
280 INPUT"(must be < 26)=";TS
290 IF AI=2 THEN 320
300 LPRINT"Number of simulation periods for which contaminant"
310 LPRINT"concentration distribution is to be calculated";TS
320 PRINT:PRINT"Time schedules for contaminant source point sources"
330 PRINT"should be prepared. Simulation period durations are equal"
340 PRINT"to the longest time schedules in each period."
350 PRINT
360 IF AI=1 THEN LPRINT
```

```
370 FOR I=1 TO TS
380 PRINT"Simulation period number=";I
390 IF AI=2 THEN 410
400 LPRINT"Simulation period number=";I
410 INPUT"Simulation period duration in days=";DELTA(I)
420 IF AI=2 THEN 440
430 LPRINT"Simulation period duration in days=" USING AT$;DELTA(I)
440 NEXT I
450 PRINT:PRINT"Contaminant point sources and other features"
460 PRINT"within the area of concern should be drawn to scale"
470 PRINT"on a map. A uniform grid should be superposed over"
480 PRINT"the map. Grid lines should be indexed using the I"
490 PRINT"(column), J (row) notation colinear with the X and"
500 PRINT"Y directions, respectively. I coordinates should"
510 PRINT"increase left to right and J coordinates should increase"
520 PRINT"top to bottom. The origin of the grid should be beyond"
530 PRINT"the upper-left corner of the map."
540 PRINT:PRINT"Press any key to continue"
550 T$=INKEY$:IF T$="" THEN 550
560 PRINT
570 INPUT"Number of grid columns (must be < 31)=";NC
580 IF AI=1 THEN LPRINT"Number of grid columns=";NC
590 INPUT"Number of grid rows (must be < 31)=";NR
600 IF AI=1 THEN LPRINT"Number of grid rows=";NR
610 INPUT"Grid spacing in ft=";GS
620 IF AI=1 THEN LPRINT"Grid spacing in ft=" USING AD$;GS
630 INPUT"X-coordinate of upper-left grid node in ft=";XGO
640 IF AI=2 THEN 660
650 LPRINT"X-coordinate of upper-left grid node in ft=" USING AD$;XGO
660 INPUT"Y-coordinate of upper-left grid node in ft=";YGO
670 IF AI=2 THEN 690
680 LPRINT"Y-coordinate of upper-left grid node in ft=" USING AD$;YGO
690 INPUT"Aquifer actual porosity as a decimal=";AP
700 IF AI=2 THEN 720
710 LPRINT"Aquifer actual porosity as a decimal=" USING AY$;AP
720 INPUT"Aquifer effective porosity as a decimal=";EP
730 IF AI=2 THEN 750
740 LPRINT"Aquifer effective porosity as a decimal=" USING AY$;EP
750 INPUT"Aquifer thickness in ft=";THICK
760 FOR M=1 TO TS
770 PRINT:PRINT"Simulation period number=";M
780 IF AI=2 THEN 810
```

```
790 LPRINT"Simulation period number=";M
800 LPRINT"Aquifer thickness in ft=" USING AK$;THICK
810 INPUT"Aquifer longitudinal dispersivity in ft=";DISPL(M)
820 IF AI=2 THEN 840
830 LPRINT"Aquifer longitudinal dispersivity in ft=" USING AK$;DISPL(M)
840 INPUT"Aquifer transverse dispersivity in ft=";DISPT(M)
850 IF AI=2 THEN 870
860 LPRINT"Aquifer transverse dispersivity in ft=" USING AK$;DISPT(M)
870 IF M > 1 THEN 910
880 PRINT:PRINT"Groundwater flow with a uniform velocity in the X-"
890 PRINT"direction is assumed. Seepage velocity is the Darcy"
900 PRINT"velocity divided by the effective porosity.":PRINT
910 INPUT"Seepage velocity in ft/day=";VEL(M)
920 IF AI=2 THEN 940
930 LPRINT"Seepage velocity in ft/day=" USING AV$;VEL(M)
940 INPUT"Number of point sources (must be < 31)=";NPTS(M)
950 IF AI=1 THEN LPRINT"Number of point sources=";NPTS(M)
960 NEXT M
970 FOR M=1 TO TS
980 PRINT:PRINT"Simulation period number=";M
990 IF AI=1 THEN LPRINT"Simulation period number=";M
1000 FOR I=1 TO NPTS(M)
1010 PRINT"Point source number=";I
1020 IF AI=1 THEN LPRINT"Point source number";I
1030 PRINT:PRINT"Enter 1 for continuous point source"
1040 INPUT"Enter 2 for slug point source";CS(I,M)
1050 IF CS(I,M) <> 1 AND CS(I,M) <> 2 THEN 1030
1060 PRINT:PRINT"Native aquifer solute concentration"
1070 PRINT"is assumed to be zero":PRINT
1080 PRINT"Enter 1 to express contaminant load as mass injection rate"
1090 PRINT"in pounds per day or as injected mass in pounds"
1100 PRINT"Enter 2 to express contaminant load in terms of injection"
1110 PRINT"rate in gpd and solute concentration in mg/l or volume of"
1120 INPUT"injected mass in gal and solute concentration in mg/l";CL(I,M)
1130 IF CL(I,M) <> 1 AND CL(I,M) <> 2 THEN 1080
1140 PRINT
1150 INPUT"X-coordinate of point source in ft=";X(I,M)
1160 IF AI=2 THEN 1180
1170 LPRINT"X-coordinate of point source in ft="; USING AD$;X(I,M)
1180 INPUT"Y-coordinate of point source in ft=";Y(I,M)
1190 IF AI=2 THEN 1210
1200 LPRINT"Y-coordinate of point source in ft=" USING AD$;Y(I,M)
```

```
1210 IF CS(I,M)=2 THEN 1340
1220 IF CL(I,M)=1 THEN 1300
1230 INPUT"Continuous point source injection rate in gpd=";Q(I,M)
1240 IF AI=2 THEN 1260
1250 LPRINT"Continuous point source injection rate in gpd=" USING AD$;Q(I,M)
1260 INPUT"Point source solute concentration in mg/l=";C(I,M)
1270 IF AI=2 THEN 1290
1280 LPRINT"Point source solute concentration in mg/l=" USING AC$;C(I,M)
1290 GOTO 1450
1300 INPUT"Contaminant injection rate in pounds/day=";CM(I,M)
1310 IF AI=2 THEN 1330
1320 LPRINT"Contaminant injection rate in pounds/day=" USING AL$;CM(I,M)
1330 GOTO 1450
1340 IF CL(I,M)=1 THEN 1420
1350 INPUT"Slug point source solute injection volume in gal=";VS(I,M)
1360 IF AI=2 THEN 1380
1370 LPRINT"Slug point source solute inject. vol. in gal=" USING AD$;VS(I,M)
1380 INPUT"Slug point source solute concentration in mg/l=";C(I,M)
1390 IF AI=2 THEN 1410
1400 LPRINT"Slug point source solute concentration in mg/l=" USING AC$;C(I,M)
1410 GOTO 1450
1420 INPUT"Slug point source contaminant load in pounds=";CM(I,M)
1430 IF AI=2 THEN 1450
1440 LPRINT"Slug point source contaminant load in pounds=" USING AL$;CM(I,M)
1450 IF CS(I,M)=2 THEN 1520
1460 PRINT"Time after continuous contaminant injection"
1470 INPUT"started in days=";TIME(I,M)
1480 IF AI=2 THEN 1510
1490 LPRINT"Time after continuous contaminant injection"
1500 LPRINT"started in days=" USING AT$;TIME(I,M)
1510 GOTO 1550
1520 INPUT"Time after slug contaminant injection in days=";TIME(I,M)
1530 IF AI=2 THEN 1550
1540 LPRINT"Time after slug contaminant injection in days=" USING AT$;TIME(I,M)
1550 NEXT I
1560 NEXT M
1570 PRINT:PRINT"In simulating adsorption, it is assumed that"
1580 PRINT"single contaminant reactions are fast and reversible,"
1590 PRINT"the isotherm is linear, and there is no mixture of"
1600 PRINT"contaminants.":PRINT
1610 PRINT:PRINT"Enter 1 to simulate adsorption during all simulation"
1620 INPUT"periods or enter 2 for no adsorption";AN
```

```
1630 IF AN <> 1 AND AN <> 2 THEN 1610
1640 PRINT
1650 PRINT"Enter 1 to simulate radioactive decay during all simulation"
1660 INPUT"periods or enter 2 for no radioactive decay";RY
1670 IF RY <> 1 AND RY <> 2 THEN 1650
1680 IF AN=2 THEN 1770
1690 PRINT
1700 INPUT"Bulk density of dry aquifer skeleton in g/cu cm=";DEN
1710 IF AI=2 THEN 1730
1720 LPRINT"Bulk density of dry aquifer skeleton in g/cu cm=" USING AK$;DEN
1730 INPUT"Aquifer distribution coefficient in ml/g=";KD
1740 IF AI=2 THEN 1760
1750 LPRINT"Aquifer distribution coefficient in ml/g=";KD
1760 RD1=1+((DEN/EP)*KD)
1770 IF RY=2 THEN 1910
1780 PRINT
1790 INPUT"Half life of radionuclide in years=";HL
1800 IF AI=2 THEN 1820
1810 LPRINT"Half life of radionuclide in years=";HL
1820 HL=HL*365
1830 PRINT:PRINT"It is assumed that radioactive decay of"
1840 PRINT"contaminants from all sources started at"
1850 PRINT"the beginning of the simulation period"
1860 PRINT
1870 Z1=.693/HL
1880 FOR M=1 TO TS
1890 Z2(M)=2.71828183#^(-Z1*DELTA(M))
1900 NEXT M
1910 PRINT:PRINT"Monitor wells are located at grid nodes":PRINT
1920 PRINT"Number of monitor wells for which time-"
1930 INPUT"concentration tables are desired (must be < 26)=";NOBS
1940 IF AI=2 THEN 1970
1950 LPRINT"Number of monitor wells for which time-"
1960 LPRINT"concentration tables are desired=";NOBS
1970 FOR I=1 TO NOBS
1980 PRINT"Monitor well number=";I
1990 IF AI=1 THEN LPRINT"Monitor well number=";I
2000 INPUT"I-coordinate of monitor well=";IO(I)
2010 IF AI=1 THEN LPRINT"I-coordinate of monitor well=";IO(I)
2020 INPUT"J-coordinate of monitor well=";JO(I)
2030 IF AI=1 THEN LPRINT"J-coordinate of monitor well=";JO(I)
2040 NEXT I
```

```
2050 PRINT
2060 FOR J=1 TO TS
2070 FOR I=1 TO NOBS
2080 CCO(I,J)=0
2090 NEXT I,J
2100 PRINT:PRINT"Enter Y to revise data base"
2110 INPUT"or N to continue program";R$
2120 IF R$ <> "Y" AND R$ <> "y" AND R$ <> "N" AND R$ <> "n" THEN 2100
2130 IF R$="Y" OR R$="y" THEN 220
2140 PRINT
2150 PRINT
2160 IF AI=1 THEN LPRINT
2170 'CALCULATE GRID NODE COORDINATES
2180 FOR J=1 TO NR
2190 CLS:PRINT"COMPUTATIONS ARE IN PROGRESS"
2200 FOR I=1 TO NC
2210 IF I=1 THEN 2240
2220 XOB(I)=GS*(I-1)+XGO
2230 GOTO 2250
2240 XOB(I)=XGO
2250 IF J=1 THEN 2280
2260 YOB(J)=GS*(J-1)+YGO
2270 GOTO 2290
2280 YOB(J)=YGO
2290 NEXT I,J
2300 'CALCULATE DISTANCES BETWEEN POINT SOURCES AND GRID
2310 'NODES, WELL FUNCTIONS, AND CONCENTRATIONS
2320 FOR M=1 TO TS
2330 IF AN=2 THEN 2370
2340 VEL(M)=VEL(M)/RD1
2350 DISPL(M)=DISPL(M)/RD1
2360 DISPT(M)=DISPT(M)/RD1
2370 FOR K=1 TO NR
2380 FOR J=1 TO NC
2390 CC(J,K)=0
2400 FOR I=1 TO NPTS(M)
2410 CLS:PRINT"COMPUTATIONS ARE IN PROGRESS"
2420 IF CS(I,M)=2 THEN 2590
2430 DX=ABS(XOB(J)-X(I,M)):DY=ABS(YOB(K)-Y(I,M))
2440 IF DX=0 THEN DX=1:IF DY=0 THEN DY=1
2450 F=DX^2+(DY^2*(DISPL(M)/DISPT(M)))
2460 S=F^.5/(2*DISPL(M))
```

```
2470 D=(XOB(J)-X(I,M))/(2*DISPL(M))
2480 IF D>50 THEN D=50
2490 U=F/(4*DISPL(M)*VEL(M)*TIME(I,M))
2500 GOSUB 4760
2510 IF CL(I,M)=1 THEN 2550
2520 ZC=.01064*C(I,M)*Q(I,M)*EXP(D)*WS
2530 ZC=ZC/(THICK*EP*(DISPL(M)*VEL(M)*DISPT(M)*VEL(M))^.5)
2540 GOTO 2570
2550 ZC=1275!*CM(I,M)*EXP(D)*WS
2560 ZC=ZC/(THICK*EP*(DISPL(M)*VEL(M)*DISPT(M)*VEL(M))^.5)
2570 CC(J,K)=CC(J,K)+ZC
2580 GOTO 2700
2590 B1=((XOB(J)-X(I,M))-VEL(M)*TIME(I,M))^2
2600 B2=4*DISPL(M)*VEL(M)*TIME(I,M)
2610 B3=(YOB(K)-Y(I,M))^2/(4*DISPT(M)*VEL(M)*TIME(I,M))
2620 B4=THICK*EP*(DISPL(M)*DISPT(M))^.5*TIME(I,M)*VEL(M)
2630 F1=(B1/B2)+B3
2640 IF F1>50 THEN F1=50
2650 IF CL(I,M)=1 THEN 2680
2660 ZC=.01064*C(I,M)*VS(I,M)*EXP(-F1)/B4
2670 GOTO 2690
2680 ZC=1275!*CM(I,M)*EXP(-F1)/B4
2690 CC(J,K)=CC(J,K)+ZC
2700 NEXT I,J,K
2710 FOR K=1 TO NR
2720 FOR J=1 TO NC
2730 IF AN=1 AND RY=1 THEN CC(J,K)=CC(J,K)*Z2(M)/RD1
2740 IF AN=1 AND RY=1 THEN 2780
2750 IF AN=1 THEN CC(J,K)=CC(J,K)/RD1
2760 IF AN=1 THEN 2780
2770 IF RY=1 THEN CC(J,K)=CC(J,K)*Z2(M)
2780 NEXT J,K
2790 FOR I=1 TO NOBS
2800 FOR K=1 TO NR
2810 FOR J=1 TO NC
2820 IF J=IO(I) AND K=JO(I) THEN CCO(I,M)=CC(J,K)
2830 NEXT J,K,I
2840 PRINT:PRINT"Nodal computation results will now be"
2850 PRINT"displayed in tabular form, use Ctrl-Num Lock"
2860 PRINT"keys to control screen scrolling"
2870 PRINT
2880 IF AI=1 THEN LPRINT
```

```
2890 PRINT"Press any key to continue"
2900 T$=INKEY$:IF T$="" THEN 2900
2910 CLS:PRINT:PRINT"NODAL COMPUTATION RESULTS:"
2920 IF AI=1 THEN LPRINT"NODAL COMPUTATION RESULTS:"
2930 PRINT
2940 IF AI=1 THEN LPRINT
2950 PRINT"SIMULATION PERIOD DURATION IN DAYS:" USING AT$;DELTA(M)
2960 IF AI=2 THEN 2980
2970 LPRINT"SIMULATION PERIOD DURATION IN DAYS:" USING AT$;DELTA(M)
2980 PRINT
2990 IF AI=1 THEN LPRINT
3000 PRINT"VALUES OF CONTAMINANT CONCENTRATION (MG/L) AT NODES:"
3010 IF AI=2 THEN 3030
3020 LPRINT"VALUES OF CONTAMINANT CONCENTRATION (MG/L) AT NODES:"
3030 PRINT
3040 IF AI=1 THEN LPRINT
3050 PRINT"J-ROW " SPC(26) "I-COLUMN"
3060 PRINT"         1      2      3       4";
3070 PRINT"     5      6      7       8":PRINT
3080 IF AI=2 THEN 3120
3090 LPRINT"J-ROW " SPC(26) "I-COLUMN"
3100 LPRINT"         1      2      3       4";
3110 LPRINT"     5      6      7       8":LPRINT
3120 IF NC=8 OR NC>8 THEN NCC=8
3130 IF NC<8 THEN NCC=NC
3140 FOR J=1 TO NR
3150 IF J>9 THEN 3180
3160 PRINT" ";J;
3170 GOTO 3190
3180 PRINT"";J;
3190 FOR I=1 TO NCC
3200 IF I=NCC THEN PRINT USING A$;CC(I,J)
3210 IF I=NCC THEN 3230
3220 PRINT USING A$;CC(I,J);
3230 NEXT I,J:PRINT
3240 IF NC<8 THEN NCC=NC
3250 IF NC=8 OR NC>8 THEN NCC=8
3260 FOR J=1 TO NR
3270 IF J>9 THEN 3300
3280 IF AI=1 THEN LPRINT" ";J;
3290 GOTO 3310
3300 IF AI=1 THEN LPRINT"";J;
```

```
3310 FOR I=1 TO NCC
3320 IF AI=2 THEN 3340
3330 IF I=NCC THEN LPRINT USING A$;CC(I,J)
3340 IF I=NCC THEN 3360
3350 IF AI=1 THEN LPRINT USING A$;CC(I,J);
3360 NEXT I,J:PRINT
3370 IF AI=1 THEN LPRINT
3380 IF NC<9 THEN 4160
3390 PRINT"J-ROW " SPC(26) "I-COLUMN"
3400 PRINT"          9        10        11        12";
3410 PRINT"        13        14        15        16":PRINT
3420 IF AI=2 THEN 3460
3430 LPRINT"J-ROW " SPC(26) "I-COLUMN"
3440 LPRINT"          9        10        11        12";
3450 LPRINT"        13        14        15        16":LPRINT
3460 IF NC<16 THEN NCC=NC
3470 IF NC=16 OR NC>16 THEN NCC=16
3480 FOR J=1 TO NR
3490 IF J>9 THEN 3530
3500 PRINT" ";J;
3510 IF AI=1 THEN LPRINT" ";J;
3520 GOTO 3550
3530 PRINT"";J;
3540 IF AI=1 THEN LPRINT"";J;
3550 FOR I=9 TO NCC
3560 IF I=NCC THEN PRINT USING A$;CC(I,J)
3570 IF AI=2 THEN 3590
3580 IF I=NCC THEN LPRINT USING A$;CC(I,J)
3590 IF I=NCC THEN 3620
3600 PRINT USING A$;CC(I,J);
3610 IF AI=1 THEN LPRINT USING A$;CC(I,J);
3620 NEXT I,J:PRINT:CLS
3630 IF AI=1 THEN LPRINT
3640 IF NC<17 THEN 4160
3650 PRINT"J-ROW " SPC(26) "I-COLUMN"
3660 PRINT"         17        18        19        20";
3670 PRINT"        21        22        23        24":PRINT
3680 IF AI=2 THEN 3720
3690 LPRINT"J-ROW " SPC(26) "I-COLUMN"
3700 LPRINT"         17        18        19        20";
3710 LPRINT"        21        22        23        24":LPRINT
3720 IF NC<24 THEN NCC=NC
```

```
3730 IF NC=24 OR NC>24 THEN NCC=24
3740 FOR J=1 TO NR
3750 IF J>9 THEN 3790
3760 PRINT" ";J;
3770 IF AI=1 THEN LPRINT" ";J;
3780 GOTO 3810
3790 PRINT"";J;
3800 IF AI=1 THEN LPRINT"";J;
3810 FOR I=17 TO NCC
3820 IF I=NCC THEN PRINT USING A$;CC(I,J)
3830 IF AI=2 THEN 3850
3840 IF I=NCC THEN LPRINT USING A$;CC(I,J)
3850 IF I=NCC THEN 3880
3860 PRINT USING A$;CC(I,J);
3870 IF AI=1 THEN LPRINT USING A$;CC(I,J);
3880 NEXT I,J:PRINT:CLS
3890 IF AI=1 THEN LPRINT
3900 IF NC<25 THEN 4160
3910 PRINT"J-ROW " SPC(26) "I-COLUMN"
3920 PRINT"        25      26      27      28";
3930 PRINT"      29      30":PRINT
3940 IF AI=2 THEN 3980
3950 LPRINT"J-ROW " SPC(26) "I-COLUMN"
3960 LPRINT"        25      26      27      28";
3970 LPRINT"      29      30":LPRINT
3980 IF NC<30 THEN NCC=NC
3990 IF NC=30 THEN NCC=30
4000 FOR J=1 TO NR
4010 IF J>9 THEN 4050
4020 PRINT" ";J;
4030 IF AI=1 THEN LPRINT" ";J;
4040 GOTO 4070
4050 PRINT"";J;
4060 IF AI=1 THEN LPRINT"";J;
4070 FOR I=25 TO NCC
4080 IF I=NCC THEN PRINT USING A$;CC(I,J)
4090 IF AI=2 THEN 4110
4100 IF I=NCC THEN LPRINT USING A$;CC(I,J)
4110 IF I=NCC THEN 4140
4120 PRINT USING A$;CC(I,J);
4130 IF AI=1 THEN LPRINT USING A$;CC(I,J);
4140 NEXT I,J
```

```
4150 IF AI=1 THEN LPRINT
4160 PRINT
4170 PRINT"Enter Y to create a sequential data file of results"
4180 PRINT"for export to graphics programs GWGRAF or OMNIPLOT."
4190 INPUT"Enter N for no file";CD$:PRINT
4200 IF CD$<>"N" AND CD$<>"n" AND CD$<>"Y" AND CD$<>"y" THEN 4170
4210 IF CD$="N" OR CD$="n" THEN 4290
4220 INPUT"Enter a valid filename with extension .DAT such as FILE1.DAT";F$
4230 OPEN F$ FOR OUTPUT AS #1
4240 FOR J=1 TO NR
4250 FOR I=1 TO NC
4260 PRINT #1, USING"####.##";CC(I,J)
4270 NEXT I,J
4280 CLOSE #1
4290 PRINT
4300 PRINT"Enter Y to create a sequential data file of results"
4310 PRINT"for export to graphics program SURFER."
4320 INPUT"Enter N for no file";CD$:PRINT
4330 IF CD$<>"N" AND CD$<>"n" AND CD$<>"Y" AND CD$<>"y" THEN 4300
4340 IF CD$="N" OR CD$="n" THEN 4420
4350 INPUT"Enter a valid filename with extension .DAT such as FILE1.DAT";F$
4360 OPEN F$ FOR OUTPUT AS #1
4370 FOR J=1 TO NR
4380 FOR I=1 TO NC
4390 WRITE #1,XOB(I),YOB(J),CC(I,J)
4400 NEXT I,J
4410 CLOSE #1
4420 NEXT M
4430 FOR J=1 TO NOBS
4440 IF J > 1 THEN 4590
4450 PRINT:PRINT"Monitor well computation results will now be"
4460 PRINT"displayed in tabular form, use Ctrl-Num Lock"
4470 PRINT"keys to control screen scrolling"
4480 PRINT
4490 IF AI=1 THEN LPRINT
4500 PRINT"Press any key to continue"
4510 T$=INKEY$:IF T$="" THEN 4510
4520 CLS:PRINT
4530 IF AI=1 THEN LPRINT
4540 PRINT"MONITOR WELL COMPUTATION RESULTS:":PRINT
4550 IF AI=1 THEN LPRINT"MONITOR WELL COMPUTATION RESULTS:"
4560 IF AI=1 THEN LPRINT
```

```
4570 PRINT"TIME-CONCENTRATION TABLE"
4580 IF AI=1 THEN LPRINT"TIME-CONCENTRATION TABLE"
4590 PRINT
4600 IF AI=1 THEN LPRINT
4610 PRINT"MONITOR WELL NUMBER:";J
4620 IF AI=1 THEN LPRINT"MONITOR WELL NUMBER:";J
4630 PRINT
4640 IF AI=1 THEN LPRINT
4650 PRINT"TIME(DAYS)      CONCENTRATION(MG/L)"
4660 IF AI=1 THEN LPRINT"TIME(DAYS)      CONCENTRATION(MG/L)"
4670 FOR I=1 TO TS
4680 PRINT USING"#####.###";DELTA(I);
4690 PRINT"        ";
4700 PRINT USING"#########.##";CCO(J,I)
4710 IF AI=1 THEN LPRINT USING"#####.###";DELTA(I);
4720 IF AI=1 THEN LPRINT"        ";
4730 IF AI=1 THEN LPRINT USING"#########.##";CCO(J,I)
4740 NEXT I,J
4750 END
4760 'SUBROUTINE TO COMPUTE W(U,S) USING POLYNOMIAL APPROXIMATIONS
4770 IF U>5! THEN WS=0:GOTO 5240
4780 IF S>2 THEN 5140
4790 IF U>=1 THEN 4820
4800 WU=-LOG(U)-.5772156600000009#+.9999919300000017#*U-.24991055#*U*U
4810 WU=WU+.0551997*U^3-9.76004E-03*U^4+1.07857E-03*U^5:GOTO 4870
4820 WU=U^4+8.573328740100025#*U^3+18.059016973#*U^2+8.634760892499999#*U
4830 WU=WU+.2677737343#
4840 WUP=U^4+9.573322345400009#*U^3+25.6329561486#*U^2+21.0996530827#*U
4850 WUP=WUP+3.95849692280001#
4860 WU=(WU/WUP)/(U*EXP(U))
4870 L=S/3.75
4880 V=1+3.5156229#*L^2+3.0899424#*L^4+1.2067492#*L^6+.265973*L^8+.0360768*L^10
4890 V=V+.0045813*L^12
4900 F=S/2
4910 H=-LOG(F)*V-.5772156600000009#+.422784*F^2+.23069756#*F^4+.0348859*F^6
4920 H=H+2.62698E-03*F^8+.0001075*F^10+.0000074*F^12
4930 IF S=0 THEN WS=WU:GOTO 5240
4940 N=S^2/(4*U)
4950 IF N>5 THEN WS=2*H:GOTO 5240
4960 IF U<=.9 THEN 4990
4970 A=U+.5858:B=U+3.414:C=S*S/4
4980 WS=1.5637*EXP(-A-C/A)/A+4.54*EXP(-B-C/B)/B:GOTO 5240
```

```
4990 IF U<.05 THEN 5030
5000 IF U>S/2 THEN 5020
5010 C=-(1.75*U)^-.448*S:WS=2*H-4.8*10^C:GOTO 5240
5020 WS=WU-(S/(4.7*U^.6))^2:GOTO 5240
5030 IF U>.01 AND S<.1 THEN 5020
5040 IF N<1 THEN 5110
5050 WN=N^4+8.573328740100025#*N^3+18.0590169737#*N^2+8.634760892499999#*N
5060 WN=WN+.267773734300001#
5070 WNP=N^4+9.573322345400009#*N^3+25.6329561486#*N^2+21.0996530827#*N
5080 WNP=WNP+3.958496922800012#
5090 WN=WN/WNP
5100 WN=WN/(N*EXP(N)):GOTO 5130
5110 WN=-LOG(N)+.999991930000017#*N-.57721566000000009#-.24991055#*N^2
5120 WN=WN+.0551997*N^3-9.76004E-03*N^4+1.07857E-03*N^5
5130 WS=2*H-WN*V:GOTO 5240
5140 MX=-((S-2*U)/(2*U^.5))
5150 IF MX<0 THEN MX=ABS(MX):GOSUB 5170:GOTO 5230
5160 GOSUB 5170:GOTO 5220
5170 NP=1+.0705230784#*MX+.0422820123#*MX^2+9.270527200000026D-03*MX^3
5180 NP=NP+1.52014E-04*MX^4+2.76567E-04*MX^5+4.30638E-05*MX^6
5190 IF NP>100! THEN 5210
5200 N=1/NP^16:RETURN
5210 N=0!:RETURN
5220 WS=(3.1416/(2*S))^.5*EXP(-S)*N:GOTO 5240
5230 WS=(3.1416/(2*S))^.5*EXP(-S)*(2-N)
5240 RETURN
```

Appendix D
GWGRAF Source Code

The source code for the GWGRAF IBM PC BASICA microcomputer program is listed in this appendix. Program algorithms are documented as REM statements in the listing to assist the reader in understanding the code. Many statements in the program could be restructured or eliminated to produce more efficient coding, but they have been intentionally written in their present form to provide clear explanations of processing steps.

```
10 KEY OFF
20 AD$="#####.##":AF$="####.##":AT$="#######.##":AS$="##.######"
30 AG$="####":AD1$="######.##"
40 DIM X(30),YK(30,30),WD(6),EL(30),XC(30),YC(30),ELL(30)
50 DIM XP(30),YP(30),Y(30),DD(30,30),C(10)
60 DIM IDIF(30),CB%(42),C1(30),C2(30),C3(30)
70 PRINT"Program: GWGRAF"
80 PRINT"Version: IBM/PC BASICA 2.1"
90 PRINT"         Copyright 1988 Lewis Publishers, Inc."
100 PRINT"         HIGH RESOLUTION MONITOR (640 BY 200)"
110 PRINT"         IBM Color Adapter or compatible device"
120 PRINT"         OPTIONAL DOT-MATRIX PRINTER"
130 PRINT"Purpose: Creating graphs and maps from"
140 PRINT"         groundwater model results"
150 PRINT"Author : William C. Walton"
160 PRINT"Date   : May, 1988"
170 PRINT:PRINT"If you do not have the required monitor"
180 PRINT"and device then a fatal error will occur"
190 PRINT:PRINT"Do you have a printer?"
200 INPUT"Enter 1 for yes or 2 for no";AI
210 IF AI <> 1 AND AI <> 2 THEN 190
220 IF AI=2 THEN 260
230 PRINT:PRINT"Turn on the printer or else an error will occur"
240 PRINT:PRINT"Press any key to continue"
250 P$=INKEY$:IF P$="" THEN 250
260 PRINT
270 PRINT:PRINT"GRAPH AND MAP OPTIONS:"
280 PRINT:PRINT"Enter 1 to create a square grid pattern of XYZ"
290 PRINT"values from scattered surface data"
300 PRINT"Enter 2 to create a time-drawdown"
310 PRINT"semilog graph with best fit line"
320 PRINT"Enter 3 to create a time-drawdown or"
330 PRINT"time-concentration arithmetic graph"
340 INPUT"Enter 4 to create a contour map of XYZ values";GF
350 IF GF<>1 AND GF<>2 AND GF<>3 AND GF<>4 THEN 280
360 IF GF=1 THEN 400
```

```
370 IF GF=2 THEN 2940
380 IF GF=3 THEN 4390
390 IF GF=4 THEN 5660
400 CLS:PRINT:PRINT"DATA BASE:":PRINT
410 IF AI=1 THEN LPRINT"DATA BASE:"
420 IF AI=1 THEN LPRINT
430 INPUT"Number of data points (must be > 5 AND < 31)=";N
440 IF AI=1 THEN LPRINT"Number of data points=";N
450 IDIF=0
460 FOR I=1 TO N
470 PRINT"Data point number=";I
480 IF AI=1 THEN LPRINT"Data point number=";I
490 INPUT"Z-value of data point in ft or mg/l=";EL(I)
500 IF AI=2 THEN 520
510 LPRINT"Z-value of data point in ft or mg/l=" USING AD1$;EL(I)
520 INPUT"X-coordinate of data point in ft=";XC(I)
530 IF AI=2 THEN 550
540 LPRINT"X-coordinate of data point in ft=" USING AD1$;XC(I)
550 INPUT"Y-coordinate of data point in ft=";YC(I)
560 IF AI=2 THEN 580
570 LPRINT"Y-coordinate of data point in ft=" USING AD1$;YC(I)
580 NEXT I
590 INPUT"Number of grid columns (must be < 31)=";NC
600 IF AI=1 THEN LPRINT"Number of grid columns=";NC
610 INPUT"Number of grid rows (must be < 31)=";NR
620 IF AI=1 THEN LPRINT"Number of grid rows=";NR
630 INPUT"Grid spacing in ft=";GP
640 IF AI=1 THEN LPRINT"Grid spacing in ft=" USING AD$;GP
650 INPUT"X-coordinate of upper-left grid node in ft=";XGO
660 IF AI=2 THEN 680
670 LPRINT"X-coordinate of upper-left grid node in ft=" USING AD1$;XGO
680 INPUT"Y-coordinate of upper-left grid node in ft=";YGO
690 IF AI=2 THEN 710
700 LPRINT"Y-coordinate of upper-left grid node in ft=" USING AD1$;YGO
710 PRINT:PRINT"Enter Y to revise data base"
720 INPUT"or N to continue";R$
730 IF R$="Y" OR R$="y" THEN 430
740 PRINT
750 PRINT"Computations may take a few minutes so be patient"
760 PRINT
770 PRINT"Press any key to continue"
780 T$=INKEY$:IF T$="" THEN 780
```

```
790 PRINT
800 FOR IJ=1 TO NR
810 FOR II=1 TO NC
820 'COMPUTE DISTANCES BETWEEN GRID NODES AND DATA POINTS
830 CLS:PRINT"COMPUTATIONS ARE IN PROGRESS"
840 IF II=1 THEN 870
850 XCG=GP*(II-1)+XGO
860 GOTO 880
870 XCG=XGO
880 IF IJ=1 THEN 910
890 YCG=GP*(IJ-1)+YGO
900 GOTO 920
910 YCG=YGO
920 FOR I=1 TO N
930 IDIF(I)=0:C1(I)=0:C2(I)=0:C3(I)=0
940 X(I)=((XCG-XC(I))^2+(YCG-YC(I))^2)^.5
950 ELL(I)=EL(I)
960 XDIF=XCG-XC(I):YDIF=YCG-YC(I)
970 IF XDIF=0 AND YDIF=0 THEN 990
980 GOTO 1000
990 IDIF(I)=1:C1(I)=II:C2(I)=IJ:C3(I)=I
1000 NEXT I
1010 FOR STP=1 TO N-1
1020 FOR J=(STP+1) TO N
1030 CLS:PRINT"COMPUTATIONS ARE IN PROGRESS"
1040 'SORT DISTANCES IN ASCENDING ORDER
1050 IF X(J)<X(STP) THEN SWAP EL(J),EL(STP)
1060 IF X(J)<X(STP) THEN SWAP X(J),X(STP)
1070 NEXT J,STP
1080 TM1=0:TM2=0:WDS=0
1090 FOR I=1 TO 6
1100 'COMPUTE WEIGHTING FACTORS FOR 6 CLOSEST DATA POINTS
1110 WD(I)=(1-(X(I)/X(6)))^2/(X(I)/X(6))
1120 WDS=WDS+WD(I)
1130 NEXT I
1140 FOR I=1 TO 6
1150 'INTERPRET Z VALUES AT GRID NODES
1160 WD(I)=WD(I)/WDS
1170 TM1=TM1+WD(I)*(EL(I)/X(1))
1180 TM2=TM2+WD(I)*(1/X(I))
1190 NEXT I
1200 YK(II,IJ)=TM1/TM2
```

```
1210 FOR I=1 TO N
1220 EL(I)=ELL(I)
1230 IF IDIF(I)=1 THEN YK(C1(I),C2(I))=EL(C3(I))
1240 NEXT I
1250 NEXT II,IJ
1260 PRINT:PRINT"Nodal computation results will now be"
1270 PRINT"displayed in tabular form, use Ctrl-Num Lock"
1280 PRINT"keys to control screen scrolling"
1290 PRINT
1300 IF AI=1 THEN LPRINT
1310 PRINT"Press any key to continue"
1320 T$=INKEY$:IF T$="" THEN 1320
1330 CLS:PRINT:PRINT"NODAL COMPUTATION RESULTS:"
1340 IF AI=1 THEN LPRINT"NODAL COMPUTATION RESULTS:"
1350 PRINT
1360 IF AI=1 THEN LPRINT
1370 PRINT"VALUES OF Z IN FT OR MG/L AT NODES:"
1380 IF AI=1 THEN LPRINT"VALUES OF Z IN FT OR MG/L AT NODES:"
1390 PRINT
1400 IF AI=1 THEN LPRINT
1410 PRINT"J-ROW " SPC(26) "I-COLUMN"
1420 PRINT"        1       2       3       4";
1430 PRINT"        5       6       7       8":PRINT
1440 IF AI=2 THEN 1480
1450 LPRINT"J-ROW " SPC(26) "I-COLUMN"
1460 LPRINT"        1       2       3       4";
1470 LPRINT"        5       6       7       8":LPRINT
1480 IF NC=8 OR NC>8 THEN NCC=8
1490 IF NC<8 THEN NCC=NC
1500 FOR J=1 TO NR
1510 IF J>9 THEN 1540
1520 PRINT" ";J;
1530 GOTO 1550
1540 PRINT"";J;
1550 FOR I=1 TO NCC
1560 IF I=NCC THEN PRINT USING AD$;YK(I,J)
1570 IF I=NCC THEN 1590
1580 PRINT USING AD$;YK(I,J);
1590 NEXT I,J:PRINT
1600 IF NC<8 THEN NCC=NC
1610 IF NC=8 OR NC>8 THEN NCC=8
1620 FOR J=1 TO NR
```

```
1630 IF J>9 THEN 1660
1640 IF AI=1 THEN LPRINT" ";J;
1650 GOTO 1670
1660 IF AI=1 THEN LPRINT"";J;
1670 FOR I=1 TO NCC
1680 IF AI=2 THEN 1700
1690 IF I=NCC THEN LPRINT USING AD$;YK(I,J)
1700 IF I=NCC THEN 1720
1710 IF AI=1 THEN LPRINT USING AD$;YK(I,J);
1720 NEXT I,J:PRINT
1730 IF AI=1 THEN LPRINT
1740 IF NC<9 THEN 2520
1750 PRINT"J-ROW " SPC(26) "I-COLUMN"
1760 PRINT"        9       10       11       12";
1770 PRINT"       13       14       15       16":PRINT
1780 IF AI=2 THEN 1820
1790 LPRINT"J-ROW " SPC(26) "I-COLUMN"
1800 LPRINT"        9       10       11       12";
1810 LPRINT"       13       14       15       16":LPRINT
1820 IF NC<16 THEN NCC=NC
1830 IF NC=16 OR NC>16 THEN NCC=16
1840 FOR J=1 TO NR
1850 IF J>9 THEN 1890
1860 PRINT" ";J;
1870 IF AI=1 THEN LPRINT" ";J;
1880 GOTO 1910
1890 PRINT"";J;
1900 IF AI=1 THEN LPRINT"";J;
1910 FOR I=9 TO NCC
1920 IF I=NCC THEN PRINT USING AD$;YK(I,J)
1930 IF AI=2 THEN 1950
1940 IF I=NCC THEN LPRINT USING AD$;YK(I,J)
1950 IF I=NCC THEN 1980
1960 PRINT USING AD$;YK(I,J);
1970 IF AI=1 THEN LPRINT USING AD$;YK(I,J);
1980 NEXT I,J:PRINT:CLS
1990 IF AI=1 THEN LPRINT
2000 IF NC<17 THEN 2520
2010 PRINT"J-ROW " SPC(26) "I-COLUMN"
2020 PRINT"       17       18       19       20";
2030 PRINT"       21       22       23       24":PRINT
2040 IF AI=2 THEN 2080
```

```
2050 LPRINT"J-ROW " SPC(26) "I-COLUMN"
2060 LPRINT"        17      18      19      20";
2070 LPRINT"      21      22      23      24":LPRINT
2080 IF NC<24 THEN NCC=NC
2090 IF NC=24 OR NC>24 THEN NCC=24
2100 FOR J=1 TO NR
2110 IF J>9 THEN 2150
2120 PRINT" ";J;
2130 IF AI=1 THEN LPRINT" ";J;
2140 GOTO 2170
2150 PRINT"";J;
2160 IF AI=1 THEN LPRINT"";J;
2170 FOR I=17 TO NCC
2180 IF I=NCC THEN PRINT USING AD$;YK(I,J)
2190 IF AI=2 THEN 2210
2200 IF I=NCC THEN LPRINT USING AD$;YK(I,J)
2210 IF I=NCC THEN 2240
2220 PRINT USING AD$;YK(I,J);
2230 IF AI=1 THEN LPRINT USING AD$;YK(I,J);
2240 NEXT I,J:PRINT:CLS
2250 IF AI=1 THEN LPRINT
2260 IF NC<25 THEN 2520
2270 PRINT"J-ROW " SPC(26) "I-COLUMN"
2280 PRINT"        25      26      27      28";
2290 PRINT"      29      30":PRINT
2300 IF AI=2 THEN 2340
2310 LPRINT"J-ROW " SPC(26) "I-COLUMN"
2320 LPRINT"        25      26      27      28";
2330 LPRINT"      29      30":LPRINT
2340 IF NC<30 THEN NCC=NC
2350 IF NC=30 THEN NCC=30
2360 FOR J=1 TO NR
2370 IF J>9 THEN 2410
2380 PRINT" ";J;
2390 IF AI=1 THEN LPRINT" ";J;
2400 GOTO 2430
2410 PRINT"";J;
2420 IF AI=1 THEN LPRINT"";J;
2430 FOR I=25 TO NCC
2440 IF I=NCC THEN PRINT USING AD$;YK(I,J)
2450 IF AI=2 THEN 2470
2460 IF I=NCC THEN LPRINT USING AD$;YK(I,J)
```

```
2470 IF I=NCC THEN 2500
2480 PRINT USING AD$;YK(I,J);
2490 IF AI=1 THEN LPRINT USING AD$;YK(I,J);
2500 NEXT I,J:CLS
2510 IF AI=1 THEN LPRINT
2520 PRINT:PRINT"Enter 1 if results are drawdowns"
2530 INPUT"Enter 2 if results are concentrations";RES
2540 IF RES <> 1 AND RES <> 2 THEN 2520
2550 IF RES=2 THEN 2600
2560 FOR J=1 TO NR
2570 FOR I=1 TO NC
2580 YK(I,J)=-YK(I,J)
2590 NEXT I,J
2600 PRINT:PRINT"Enter Y to create a sequential data file of results"
2610 PRINT"for export to graphics programs GWGRAF or OMNIPLOT."
2620 INPUT"Enter N for no file";CD$:PRINT
2630 IF CD$<>"N" AND CD$<>"n" AND CD$<>"Y" AND CD$<>"y" THEN 2600
2640 IF CD$="N" OR CD$="n" THEN 2720
2650 INPUT"Enter a valid filename with extension .DAT such as FILE1.DAT";F$
2660 OPEN F$ FOR OUTPUT AS #1
2670 FOR J=1 TO NR
2680 FOR I=1 TO NC
2690 PRINT #1, USING"####.##";YK(I,J)
2700 NEXT I,J
2710 CLOSE #1
2720 PRINT
2730 PRINT"Enter Y to create a sequential data file of results"
2740 PRINT"for export to graphics program SURFER."
2750 INPUT"Enter N for no file";CD$:PRINT
2760 IF CD$<>"N" AND CD$<>"n" AND CD$<>"Y" AND CD$<>"y" THEN 2730
2770 IF CD$="N" OR CD$="n" THEN 2930
2780 INPUT"Enter a valid filename with extension .DAT such as FILE1.DAT";F$
2790 OPEN F$ FOR OUTPUT AS #1
2800 FOR J=1 TO NR
2810 FOR I=1 TO NC
2820 IF I=1 THEN 2850
2830 XCG=GP*(I-1)+XGO
2840 GOTO 2860
2850 XCG=XGO
2860 IF J=1 THEN 2890
2870 YCG=GP*(J-1)+YGO
2880 GOTO 2900
```

```
2890 YCG=YGO
2900 WRITE #1,XCG,YCG,YK(I,J)
2910 NEXT I,J
2920 CLOSE #1
2930 END
2940 CLS:PRINT:PRINT"It is assumed that you entered the DOS command"
2950 PRINT"GRAPHICS so that a hard copy of the screen "
2960 PRINT"can be made as part of the program by holding"
2970 PRINT"down the Shift key and pressing the PrtSc key."
2980 PRINT"After obtaining the hard copy, press"
2990 PRINT"any key to continue.":PRINT
3000 PRINT"Drawdown (ft) is plotted on the vertical axis and"
3010 PRINT"log of the time after pumping started (min) is plotted"
3020 PRINT"on the horizontal axis."
3030 PRINT"Horizontal axis is labeled 1,10,100,1000,10000,100000."
3040 PRINT"Vertical axis will be labeled starting with 0 and"
3050 PRINT"ending with a multiple of 10 such as 10,30,100, or 500"
3060 PRINT"user-defined as the maximum number on vertical axis."
3070 INPUT"Enter maximum number on vertical axis (must be => 10)";MVN
3080 SV=MVN/10
3090 PRINT:PRINT"DATA BASE:":PRINT
3100 IF AI=1 THEN LPRINT
3110 IF AI=1 THEN LPRINT"DATA BASE:"
3120 IF AI=1 THEN LPRINT
3130 INPUT"Production well discharge rate in gpm=";Q
3140 IF AI=2 THEN 3160
3150 LPRINT"Production well discharge rate in gpm=" USING AD$;Q
3160 INPUT"Distance between prod. and obs. wells in ft=";DIST
3170 IF AI=2 THEN 3190
3180 LPRINT"Distance between prod. and obs. wells in ft=" USING AD$;DIST
3190 INPUT"Number of known points (must < 31)";N
3200 D=0
3210 D1=0
3220 D2=0
3230 D3=0
3240 PRINT:PRINT"Enter time-drawdown data in ascending time order"
3250 PRINT
3260 FOR I=1 TO N
3270 PRINT"Point number=";I
3280 INPUT"Time-coordinate of point in min=";X1
3290 INPUT"Drawdown-coordinate of point in ft=";Y1
3300 IF AI=2 THEN 3340
```

```
3310 LPRINT"Point number=";I
3320 LPRINT"Time-coordinate of point in min=";USING AD$;X1
3330 LPRINT"Drawdown-coordinate of point in ft=" USING AD$;Y1
3340 'LINEAR REGRESSION WITH METHOD OF LEAST SQUARES
3350 XP(I)=X1:YP(I)=Y1
3360 X(I)=84.5+LOG(XP(I))/LOG(10)*96
3370 Y(I)=12.5+YP(I)*16/SV
3380 X1=LOG(X1)/LOG(10)
3390 D=D+X1:D1=D1+Y1
3400 D2=D2+X1^2:D3=D3+Y1^2
3410 XY=XY+X1*Y1
3420 NEXT I
3430 D4=D/N:D5=D1/N
3440 D6=D2-D^2/N
3450 D7=D3-D1^2/N
3460 D8=(D6/N)^.5
3470 D9=(D7/N)^.5
3480 D10=XY-D*D1/N
3490 D11=D10/(D6*D7)^.5
3500 M=D11*D9/D8:B=D5-M*D4
3510 PRINT:PRINT"Enter Y to revise data base"
3520 INPUT"or N to continue";R$:PRINT
3530 IF R$<>"Y" AND R$<>"y" AND R$<>"N" AND R$<>"n" THEN 3510
3540 IF R$="Y" OR R$="y" THEN 2940
3550 'INTERPOLATION USING BEST-FIT STRAIGHT LINE EQUATION
3560 XS1=X(1)
3570 YS1=B+M*LOG(XP(1))/LOG(10)
3580 S1=YS1
3590 YS1=12.5+YS1*16/SV
3600 XS2=X(N)
3610 YS2=B+M*LOG(XP(N))/LOG(10)
3620 S2=YS2
3630 YS2=12.5+YS2*16/SV
3640 D20=XP(N)/XP(1):D30=S2-S1
3650 TRANS=264*Q*.4342945*LOG(D20)/D30
3660 PRINT
3670 IF AI=1 THEN LPRINT
3680 PRINT"Aquifer transmissivity in gpd/ft=" USING AT$;TRANS
3690 IF AI=2 THEN 3710
3700 LPRINT"Aquifer transmissivity in gpd/ft=" USING AT$;TRANS
3710 D40=S2*TRANS/(114.6*Q)+.5772
3720 STOR=TRANS*XP(N)/(2693*DIST^2*2.71828183/^D40)
```

```
3730 PRINT"Aquifer storativity as a decimal=" USING AS$;STOR
3740 IF AI=2 THEN 3760
3750 LPRINT"Aquifer storativity as a decimal=" USING AS$;STOR
3760 PRINT:PRINT"Press any key to continue"
3770 T$=INKEY$:IF T$="" THEN 3770
3780 SCREEN 2:LOCATE ,,0:CLS
3790 'LABEL GRAPH AXES AND TICKS
3800 LOCATE 1,1
3810 PRINT"DRAWDOWN(FT)"
3820 LOCATE 1,26
3830 PRINT"SEMI-LOG TIME-DRAWDOWN GRAPH"
3840 LOCATE 2,2
3850 PRINT USING AF$;0*MVN
3860 LOCATE 4,2
3870 PRINT USING AF$;.1*MVN
3880 LOCATE 6,2
3890 PRINT USING AF$;.2*MVN
3900 LOCATE 8,2
3910 PRINT USING AF$;.3*MVN
3920 LOCATE 10,2
3930 PRINT USING AF$;.4*MVN
3940 LOCATE 12,2
3950 PRINT USING AF$;.5*MVN
3960 LOCATE 14,2
3970 PRINT USING AF$;.6*MVN
3980 LOCATE 16,2
3990 PRINT USING AF$;.7*MVN
4000 LOCATE 18,2
4010 PRINT USING AF$;.8*MVN
4020 LOCATE 20,2
4030 PRINT USING AF$;.9*MVN
4040 LOCATE 22,2
4050 PRINT USING AF$;MVN
4060 LOCATE 23,10
4070 PRINT" 1        10        100       1000      10000     100000"
4080 'DRAW FRAME OF GRAPH
4090 LOCATE 22,74:PRINT"TIME"
4100 LOCATE 23,73:PRINT" (MIN)"
4110 LINE (84,12)-(564,12)
4120 LINE -(564,172)
4130 LINE -(84,172)
4140 LINE -(84,12)
```

```
4150 LINE (85,13)-(85,171):LINE (563,13)-(563,171)
4160 'DRAW GRAPH TICKS
4170 FOR I=0 TO 8
4180 YT1=28+16*I
4190 LINE (86,YT1)-(91,YT1)
4200 LINE (558,YT1)-(563,YT1)
4210 NEXT I
4220 FOR I=0 TO 3
4230 TY2=180+96*I
4240 LINE (TY2,171)-(TY2,166)
4250 LINE (TY2+1,171)-(TY2+1,166)
4260 LINE (TY2,13)-(TY2,18)
4270 LINE (TY2+1,13)-(TY2+1,18)
4280 NEXT I
4290 'DRAW SYMBOLS ON GRAPH AT DATA POINTS
4300 FOR I=1 TO N
4310 XL1=X(I)-2:XL2=X(I)+2:YL1=Y(I)-2:YL2=Y(I)+2
4320 LINE (XL1,Y(I))-(XL2,Y(I))
4330 LINE (X(I),YL1)-(X(I),YL2)
4340 NEXT I
4350 'DRAW BEST-FIT STRAIGHT LINE
4360 LINE (XS1,YS1)-(XS2,YS2)
4370 S$=INKEY$:IF S$="" THEN 4370
4380 END
4390 CLS:PRINT:PRINT"It is assumed that you entered the DOS command"
4400 PRINT"GRAPHICS so that a hard copy of the screen "
4410 PRINT"can be made as part of the program by holding"
4420 PRINT"down the Shift key and pressing the PrtSc key."
4430 PRINT"After obtaining the hard copy, press"
4440 PRINT"any key to continue.":PRINT
4450 PRINT:PRINT"Press any key to continue"
4460 T$=INKEY$:IF T$="" THEN 4460
4470 PRINT:PRINT"Enter 1 for time-drawdown graph"
4480 INPUT"Enter 2 for time concentration graph";GR
4490 IF GR<>1 AND GR<>2 THEN 4470
4500 PRINT
4510 IF GR=2 THEN 4550
4520 PRINT:PRINT"Drawdown (ft) is plotted on the vertical axis and time"
4530 PRINT"after pumping started is plotted on the horizontal axis."
4540 GOTO 4580
4550 PRINT:PRINT"Concentration (mg/l) is plotted on the vertical axis and"
4560 PRINT"time after contaminant injection is plotted on the"
```

```
4570 PRINT"horizontal axis."
4580 PRINT"Horizontal axis will be labeled starting with 0 and"
4590 PRINT"ending with a multiple of 10 such as 10,50,100,400,1000,5000"
4600 PRINT"user-defined as the maximum number on horizontal axis."
4610 PRINT:INPUT"Enter maximum number on horizontal axis (must be => 10)";MHN
4620 SH=MHN/10:PRINT
4630 PRINT"Vertical axis will be labeled starting with 0 and"
4640 PRINT"ending with a multiple of 10 such as 10,30,100, or 1000"
4650 PRINT"user-defined as the maximum number on vertical axis."
4660 PRINT:INPUT"Enter maximum number on vertical axis (must be => 10)";MVN
4670 SV=MVN/10:PRINT
4680 INPUT"Number of known points (must be < 31)";N
4690 IF AI=1 THEN LPRINT"Number of known points=";N
4700 PRINT:INPUT"Enter days or min to label horizontal time axis";TC$
4710 PRINT:PRINT"Enter data base in ascending time order":PRINT
4720 FOR I=1 TO N
4730 PRINT"Point number=";I
4740 INPUT"Time-coordinate of point in days or min=";X1
4750 INPUT"Drawdown or concentration coordinate of point in ft or mg/l=";Y1
4760 IF AI=2 THEN 4810
4770 LPRINT"Point number=";I
4780 LPRINT"Time-coordinate of point in days or min=" USING AD$;X1
4790 LPRINT"Drawdown or concentration coordinate"
4800 LPRINT"of point in ft or mg/l=" USING AD$;Y1
4810 XP(I)=X1:YP(I)=Y1
4820 X(I)=84.5+XP(I)*48/SH
4830 Y(I)=12.5+YP(I)*16/SV
4840 NEXT I
4850 PRINT:PRINT"Enter Y to revise data base"
4860 INPUT"or N to continue";R$
4870 IF R$<>"Y" AND R$<>"y" AND R$<>"N" AND R$<>"n" THEN 4850
4880 IF R$="Y" OR R$="y" THEN 4520
4890 SCREEN 2:LOCATE ,,0:CLS
4900 LOCATE 1,1
4910 IF GR=2 THEN 4940
4920 PRINT"DRAWDOWN(FT)"
4930 GOTO 4950
4940 PRINT"CONCENTRATION(MG/L)"
4950 LOCATE 1,26
4960 IF GR=2 THEN 4990
4970 PRINT"ARITHMETIC TIME-DRAWDOWN GRAPH"
4980 GOTO 5000
```

```
4990 PRINT"ARITHMETIC TIME-CONCENTRATION GRAPH"
5000 LOCATE 2,2
5010 PRINT USING AF$;0*MVN
5020 LOCATE 4,2
5030 PRINT USING AF$;.1*MVN
5040 LOCATE 6,2
5050 PRINT USING AF$;.2*MVN
5060 LOCATE 8,2
5070 PRINT USING AF$;.3*MVN
5080 LOCATE 10,2
5090 PRINT USING AF$;.4*MVN
5100 LOCATE 12,2
5110 PRINT USING AF$;.5*MVN
5120 LOCATE 14,2
5130 PRINT USING AF$;.6*MVN
5140 LOCATE 16,2
5150 PRINT USING AF$;.7*MVN
5160 LOCATE 18,2
5170 PRINT USING AF$;.8*MVN
5180 LOCATE 20,2
5190 PRINT USING AF$;.9*MVN
5200 LOCATE 22,2
5210 PRINT USING AF$;MVN
5220 LOCATE 22,74:PRINT"TIME"
5230 LOCATE 23,8:PRINT USING AG$;0*MHN
5240 LOCATE 23,14:PRINT USING AG$;.1*MHN
5250 LOCATE 23,20:PRINT USING AG$;.2*MHN
5260 LOCATE 23,26:PRINT USING AG$;.3*MHN
5270 LOCATE 23,32:PRINT USING AG$;.4*MHN
5280 LOCATE 23,38:PRINT USING AG$;.5*MHN
5290 LOCATE 23,44:PRINT USING AG$;.6*MHN
5300 LOCATE 23,50:PRINT USING AG$;.7*MHN
5310 LOCATE 23,56:PRINT USING AG$;.8*MHN
5320 LOCATE 23,62:PRINT USING AG$;.9*MHN
5330 LOCATE 23,68:PRINT USING AG$;MHN
5340 LOCATE 23,74:PRINT TC$
5350 'DRAW GRAPH FRAME
5360 LINE (84,12)-(564,12)
5370 LINE -(564,172)
5380 LINE -(84,172)
5390 LINE -(84,12)
5400 LINE (85,13)-(85,171):LINE (563,13)-(563,171)
```

```
5410 'DRAW TICKS
5420 FOR I=0 TO 8
5430 YT1=28+16*I
5440 LINE (86,YT1)-(91,YT1)
5450 LINE (558,YT1)-(563,YT1)
5460 NEXT I
5470 FOR I=0 TO 8
5480 TY2=132+48*I
5490 LINE (TY2,171)-(TY2,166)
5500 LINE (TY2+1,171)-(TY2+1,166)
5510 LINE (TY2,13)-(TY2,18)
5520 LINE (TY2+1,13)-(TY2+1,18)
5530 NEXT I
5540 'DRAW SYMBOLS ON GRAPH AT DATA POINTS
5550 FOR I=1 TO N
5560 XL1=X(I)-2:XL2=X(I)+2:YL1=Y(I)-2:YL2=Y(I)+2
5570 LINE (XL1,Y(I))-(XL2,Y(I))
5580 LINE (X(I),YL1)-(X(I),YL2)
5590 NEXT I
5600 'DRAW LINES CONNECTING POINTS
5610 FOR I=1 TO N-1
5620 LINE (X(I),Y(I))-(X(I+1),Y(I+1))
5630 NEXT I
5640 S$=INKEY$:IF S$="" THEN 5640
5650 END
5660 CLS:PRINT:PRINT"It is assumed that you used the DOS command"
5670 PRINT"GRAPHICS so that a hard copy of the screen"
5680 PRINT"can be made as part of the program by holding down"
5690 PRINT"the Shift key and pressing the PrtSc key.":PRINT
5700 PRINT"Press any key to continue"
5710 T$=INKEY$:IF T$="" THEN 5710
5720 PRINT
5730 PRINT"After a BEEP, you may label the contours"
5740 PRINT"interactively. You will see the cursor in"
5750 PRINT"the upper left corner of the grid. You may"
5760 PRINT"move the cursor to location of label lower left"
5770 PRINT"corner by pressing U for up,D for down,R"
5780 PRINT"for right,or L for left. For rapid movement,"
5790 PRINT"hold selected key down for awhile. The cursor"
5800 PRINT"must be moved for each contour label. When"
5810 PRINT"cursor is at selected location,press N then enter"
5820 PRINT:PRINT"Press any key to continue"
```

```
5830 T$=INKEY$:IF T$="" THEN 5830
5840 PRINT
5850 PRINT"the number you wish printed. The number will"
5860 PRINT"will appear briefly in the lower right portion"
5870 PRINT"of the screen and be automatically erased."
5880 PRINT"The number is then displayed as the contour label."
5890 PRINT"When you have completed contour labeling and obtained"
5900 PRINT"a hard copy of screen, press E to end program."
5910 IF AI=2 THEN 5980
5920 PRINT
5930 PRINT"Enter Y for hard copy of these instructions"
5940 INPUT"or N to continue";HI$:PRINT
5950 IF HI$<>"Y" AND HI$<>"y" AND HI$<>"N" AND HI$<>"n" THEN 5930
5960 PRINT
5970 IF HI$="Y" OR HI$="y" THEN GOSUB 7920
5980 PRINT"It is assumed that the data base to be"
5990 PRINT"contoured is in the form of a square grid"
6000 PRINT"I,J array.The data base may be filed"
6010 PRINT"in conjuction with a program that generates"
6020 PRINT"drawdown or concentration values. The format"
6030 PRINT"of the data base must be as follows: Z(1,1),"
6040 PRINT"Z(2,1),... up to Z(NC,1) where NC=number of"
6050 PRINT"columns,then,Z(1,2),Z(2,2),... up to Z(NC,2)"
6060 PRINT"These entries continue through Z(NC,NR) where"
6070 PRINT"NR=number of rows."
6080 PRINT"Enter valid data base file name"
6090 INPUT"such as FILE1.DAT";DB$:PRINT
6100 PRINT:PRINT"Number of columns and rows must"
6110 PRINT"match data base and be 10, 20, or 30."
6120 INPUT"Number of grid columns=";NC
6130 IF AI=1 THEN LPRINT"Number of grid columns=";NC
6140 INPUT"Number of grid rows=";NR
6150 IF AI=1 THEN LPRINT"Number of grid rows=";NR
6160 SIJ=NC/10:IF NR>NC THEN SIJ=NR/10
6170 INPUT"Number of contours to be printed (must be <11)";CN
6180 IF AI=1 THEN LPRINT"Number of contours to be printed=";CN
6190 S2=.04
6200 S3=.02
6210 PRINT:PRINT"Enter contour values in order from lowest to highest":PRINT
6220 FOR I=1 TO CN
6230 PRINT"Contour number=";I
6240 IF AI=1 THEN LPRINT"Contour number=";I
```

```
6250 INPUT"Contour value in ft or mg/l (must be 5 or less digits)=";C(I)
6260 IF AI=1 THEN LPRINT"Contour value in ft or mg/l=" USING AD$;C(I)
6270 NEXT I
6280 INPUT"Grid spacing in ft=";GS
6290 IF AI=1 THEN LPRINT"Grid spacing in ft=" USING AD$;GS
6300 PRINT"Enter 1 for drawdown data base"
6310 INPUT"Enter 2 for contaminant concentration data base";DB
6320 IF DB<>1 AND DB<>2 THEN 6300
6330 IF DB=1 THEN CD$="FT"
6340 IF DB=2 THEN CD$="FT"
6350 PRINT:PRINT"Enter Y to revise data base"
6360 INPUT"or N to continue";R$
6370 IF R$<>"Y" AND R$<>"y" AND R$<>"N" AND R$<>"n" THEN 6350
6380 IF R$="Y" OR R$="y" THEN 5980
6390 PRINT:PRINT"It takes about 10 minutes to draw one contour"
6400 PRINT"if program is not compiled so be patient"
6410 PRINT:PRINT"Press any key to continue"
6420 T$=INKEY$:IF T$="" THEN 6420
6430 OPEN DB$ FOR INPUT AS #1
6440 FOR J=1 TO NR
6450 FOR I=1 TO NC
6460 INPUT #1,DD(I,J)
6470 NEXT I,J
6480 FOR J=1 TO NR
6490 FOR I=1 TO NC
6500 IF DD(I,J)<0 THEN DD(I,J)=-DD(I,J)
6510 NEXT I,J
6520 CLOSE #1
6530 SCREEN 2:LOCATE ,,0:CLS
6540 AF1$="##"
6550 'LABEL GRAPH AXES AND TICKS
6560 LOCATE 3,10:PRINT"J-ROW"
6570 IF DB=2 THEN 6600
6580 LOCATE 3,20:PRINT"DRAWDOWN CONTOUR MAP"
6590 GOTO 6610
6600 LOCATE 3,17:PRINT"CONTAMINANT CONCENTRATION CONTOUR MAP"
6610 FOR II=1 TO 11
6620 IF II=1 THEN 6700
6630 I=(II-1)*2+2
6640 IG=(II-1)*SIJ
6650 LOCATE I,12
6660 PRINT USING AF1$;IG
```

```
6670 J=(II-1)*4*1.2+10
6680 LOCATE 23,J
6690 PRINT USING AF1$;IG
6700 NEXT II
6710 LOCATE 23,63:PRINT"I-COLUMN"
6720 LOCATE 3,63:PRINT"GRID SPACING="
6730 LOCATE 4,63:PRINT GS*SIJ;CD$
6740 'DRAW FRAME OF GRAPH
6750 LINE (118,28)-(470,28)
6760 LINE -(470,172)
6770 LINE -(118,172)
6780 LINE -(118,28)
6790 LINE (119,29)-(119,171):LINE (469,29)-(469,171)
6800 'DRAW TICKS
6810 FOR I=0 TO 8
6820 YT1=28+16*I
6830 LINE (121,YT1)-(126,YT1)
6840 LINE (464,YT1)-(469,YT1)
6850 NEXT I
6860 FOR I=0 TO 8
6870 TY2=118+38.4*I
6880 LINE (TY2,171)-(TY2,166)
6890 LINE (TY2+1,171)-(TY2+1,166)
6900 LINE (TY2,29)-(TY2,34)
6910 LINE (TY2+1,29)-(TY2+1,34)
6920 NEXT I
6930 'INTERPOLATE AND PLOT POINTS ON CONTOURS
6940 FOR L=1 TO CN
6950 J=1
6960 FOR I=1 TO NC-1
6970 IF INT(J)-J=0 THEN 7020
6980 J1=INT(J):J2=J1+1
6990 D1=((DD(I,J2)-DD(I,J1))*(J-J1))+DD(I,J1)
7000 D2=((DD(I+1,J2)-DD(I+1,J1))*(J-J1))+DD(I+1,J1)
7010 GOTO 7030
7020 D1=DD(I,J):D2=DD(I+1,J)
7030 IF C(L)>D1 AND C(L)<D2 THEN 7060
7040 IF C(L)<D1 AND C(L)>D2 THEN 7060
7050 GOTO 7080
7060 XC=I+(C(L)-D1)/(D2-D1)*((I+1)-I)
7070 PSET(82+XC*38.4/SIJ,12+J*16/SIJ)
7080 NEXT I
```

```
7090 J=J+S2
7100 IF J>NR THEN 7120
7110 GOTO 6960
7120 I=1
7130 FOR J=1 TO NR-1
7140 IF INT(I)-I=0 THEN 7190
7150 I1=INT(I):I2=I1+1
7160 D1=((DD(I2,J)-DD(I1,J))*(I-I1))+DD(I1,J)
7170 D2=((DD(I2,J+1)-DD(I1,J+1))*(I-I1))+DD(I1,J+1)
7180 GOTO 7200
7190 D1=DD(I,J):D2=DD(I,J+1)
7200 IF C(L)>D1 AND C(L)<D2 THEN 7230
7210 IF C(L)<D1 AND C(L)>D2 THEN 7230
7220 GOTO 7250
7230 YC=J+(C(L)-D1)/(D2-D1)*((J+1)-J)
7240 PSET(82+I*38.4/SIJ,12+YC*16/SIJ)
7250 NEXT J
7260 I=I+S3
7270 IF I>NC THEN 7290
7280 GOTO 7130
7290 NEXT L
7300 BEEP:BEEP:BEEP
7310 GOSUB 7350
7320 S$=INKEY$:IF S$="" THEN 7320
7330 END
7340 'SUBROUTINE TO WRITE CONTOUR LABELS
7350 CUR1=1.8:CUR2=.9000001
7360 X=5:Y=5:CURX=86:CURY=56
7370 PSET(CUR1*CURX,CUR2*CURY)
7380 'CURSOR CONTROL
7390 CUR3=7:CUR4=4
7400 A$=INKEY$:IF A$="" THEN 7400
7410 'LABEL CONTOURS
7420 IF A$="L" OR A$="l" THEN GOSUB 7490
7430 IF A$="R" OR A$="r" THEN GOSUB 7580
7440 IF A$="U" OR A$="u" THEN GOSUB 7660
7450 IF A$="D" OR A$="d" THEN GOSUB 7740
7460 IF A$="N" OR A$="n" THEN GOSUB 7830
7470 IF A$="E" OR A$="e" THEN 7320
7480 GOTO 7400
7490 X=CURX-CUR3:Y=CURY
7500 'SUBROUTINES FOR CURSOR MOVEMENT
```

```
7510 IF X<1 THEN 7570
7520 IF Y<1 THEN 7570
7530 IF Y>184 THEN 7570
7540 PSET(CUR1*CURX,CUR2*CURY),Ø
7550 PSET(CUR1*X,CUR2*Y)
7560 CURX=X
7570 RETURN
7580 X=CURX+CUR3:Y=CURY
7590 IF X>329 THEN 7650
7600 IF Y<1 THEN 7650
7610 IF Y>184 THEN 7650
7620 PSET(CUR1*CURX,CUR2*CURY),Ø
7630 PSET(CUR1*X,CUR2*Y)
7640 CURX=X
7650 RETURN
7660 Y=CURY-CUR4:X=CURX
7670 IF Y<1 THEN 7730
7680 IF X<1 THEN 7730
7690 IF X>624 THEN 7730
7700 PSET(CUR1*CURX,CUR2*CURY),Ø
7710 PSET(CUR1*X,CUR2*Y)
7720 CURY=Y
7730 RETURN
7740 Y=CURY+CUR4:X=CURX
7750 IF Y>184 THEN 7810
7760 IF X<1 THEN 7810
7770 IF X>624 THEN 7810
7780 PSET(CUR1*CURX,CUR2*CURY),Ø
7790 PSET(CUR1*X,CUR2*Y)
7800 CURY=Y
7810 RETURN
7820 'SUBROUTINE FOR GENERATING NUMBERS USING GET AND PUT
7830 LOCATE 18,64:INPUT"No.";CT
7840 LOCATE 16,64:PRINT CT
7850 GET(500,118)-(550,127),CB%
7860 XCURX=CUR1*CURX-10:YCURY=CUR2*CURY-10
7870 PUT (XCURX,YCURY),CB%,PSET
7880 LOCATE 18,64:PRINT"          "
7890 LOCATE 16,64:PRINT"          "
7900 RETURN
7910 IF AI=2 THEN 8110
7920 LPRINT"It is assumed that you entered the DOS command"
```

```
7930 LPRINT"GRAPHICS so that a hard copy of the screen"
7940 LPRINT"can be made as part of the program by holding down"
7950 LPRINT"the key Shift and pressing the key PrtSc."
7960 LPRINT"After a BEEP, you may label the contours"
7970 LPRINT"interactively. You will see the cursor in"
7980 LPRINT"the upper left corner of the grid. You may"
7990 LPRINT"move the cursor to location of label lower left"
8000 LPRINT"corner by pressing U for up,D for down,R"
8010 LPRINT"for right,or L for left. For rapid movement,"
8020 LPRINT"hold selected key down for awhile. The cursor"
8030 LPRINT"must be moved for each contour label. When"
8040 LPRINT"cursor is at selected location,press N then"
8050 LPRINT"the number you wish printed. The number will"
8060 LPRINT"briefly appear in the lower right portion of the"
8070 LPRINT"screen and be automatically erased."
8080 LPRINT"The number is displayed as the contour label."
8090 LPRINT"When you have completed contour labeling and obtained"
8100 LPRINT"a hard copy of the screen, press E to end program"
8110 RETURN
```

Appendix E
Diskette Instructions

Instructions for using the diskettes containing four microcomputer programs are given in this appendix. It is assumed that the reader is familiar with the operation of the microcomputer and its peripherals, and understands the Disk Operating System manual supplied with the microcomputer. Instructions assume you have two diskette drives; if you have one diskette drive or a fixed disk, consult your Disk Operating System manual. Please note that the GWGRAF program will not work with a Hercules monochrome monitor adapter card.

Basic requirements are an IBM-PC or compatible computer with 256K RAM; a high resolution (640 x 200) monitor (monochrome mode); an IBM Color Graphics Adapter (CGA), IBM Video Graphics Array (VGA), or IBM Multi-Color Graphics Array (MCGA); Hercules Color card (not Hercules *Graphics* card) or equivalent; a dot matrix printer (optional); and IBM-PC DOS or MS DOS Version 2.1 or higher.

The diskettes do not contain DOS commands because of copyright restrictions prohibiting the distribution of bootable diskettes. Insert your DOS system diskette in drive A, insert the diskettes one at a time in drive B, and turn on the power switch. Respond to prompts by entering the date and time. In response to the system prompt A > , request directories of Diskette 1 and Diskette 2, one at a time, by using the DOS command: dir b:. Your diskettes should display the following files (though not necessarily in this order): Diskette 1 WELFUN.BAS, WELFLO.BAS, CONMIG.BAS, and GWGRAF.BAS; Diskette 2

WELFUN.EXE, WELFLO.EXE, CONMIG.EXE, and GWGRAF.EXE. The stand-alone executable files were created with the QuickBASIC version 4.0 Compiler (trademark of Microsoft Corp.). If all of these files are not on the diskettes, contact the publisher.

With your DOS system diskette in drive A, remove the program diskette from drive B and insert a blank diskette in drive B. In response to the system prompt A>, format two blank diskettes using the DOS command: format b:/s. Remove your DOS system diskette from drive A and insert Diskette 1 in drive A. In response to the system prompt A>, copy Diskette 1 programs to one of the two formatted working diskettes in drive B using the following DOS command: copy *.* b:. Remove Diskette 1 from drive A and insert your DOS system diskette in drive A. In response to the system prompt A>, copy the BASICA file onto the formatted working Diskette 1 in drive B using the DOS command: copy basica.com b:. In response to the second system prompt A>, copy the GRAPHICS file onto the formatted working Diskette 1 in drive B using the DOS command: copy graphics.com b:. If you wish to use the text line editor called EDLIN, then in response to the third DOS prompt A>, copy the EDLIN file onto the formatted working Diskette 1 in drive B by using the DOS command: copy edlin.com b:. One of the two formatted working diskette's directory is now complete. To use the BASICA programs on the working Diskette 1 in drive A, in response to the system prompt A>, enter the command graphics; in response to the system prompt displayed for the second time, enter the command basica. In response to the BASICA prompt ok, enter the load command with the desired program filename such as: load "b:welfun",r and the specified program will be loaded and run.

Remove your DOS system diskette from drive A and insert Diskette 2 in drive A. In response to the system prompt A>, copy Diskette 2 programs to the formatted working Diskette 2 in drive B using the following DOS command: copy *.* b:. Remove Diskette 2 from drive A and insert your DOS system diskette in drive A. In response to the system prompt A> , copy the GRAPHICS file onto the working Diskette 2 in drive B using the DOS command: copy graphics.com b:. To use the compiled executable files on the working diskette in drive A, in response to the system prompt A> , enter the command GRAPHICS; in response to the system prompt displayed for the second time enter the base name of the desired program and the specified program will be loaded and run. If you have a printer, turn it on or else an error will occur.

The user is interactively prompted as to when and what data should be entered with specified units. The prompt is in the form of a sentence written on the display screen. Output information is displayed on the screen and at the dot matrix printer through screen dumps. Pre- or post-processor programs are not required.

The author has taken due care in preparing the diskettes, including research and testing to ensure their accuracy and effectiveness. Neither the author nor the publisher makes any warranty of any kind, expressed or implied, with regard to the performance of the diskettes and associated source codes. No warranties, expressed or implied, are made by the author or publisher that the programs are free of error, or are consistent with any particular standard of merchantability, or that they meet the reader's requirements for any particular application. In no event shall the author or publisher be liable for incidental or consequential damages in connection with or arising from the furnishing, performance, or use of the programs on the diskettes. Programs will give meaningful results only for reasonable problems. Support for the diskettes and associated source codes is this book and is not otherwise available via telephone or mail.

Although this book and its programs are copyrighted, the reader is authorized to make one machine-readable copy of each program for personal use. Distribution of the machine-readable programs (either as copied by the reader or supplied with this book) is not authorized. Before use,

programs should be verified with respect to known probable solutions over the range of available data.

The contents of this book, including the Preface should
be read and understood before programs are run. If this is
your first experience with analytical groundwater modeling, read Walton (1988) before running programs. Program
operating information is presented in Chapters 2, 4, and 6
and is displayed on the screen (see information statements
in Appendixes A-D). Several hours or days may be required
before the user becomes familiar with program operation.
The user should solve the simple problems described in
Appendix H before attempting to solve other problems.
The diskettes included with this book should not be
thought of as commercial software, but rather as aids to
learning about analytical groundwater modeling. Source
codes are provided so that the user may modify programs
as required. Particular applications are the responsibility of
the user, not the author or the publisher. If this is your first
experience with BASIC, read Weinman and Kurshan (1985)
before attempting to modify the source codes.

Appendix F
Representative Aquifer and Contamination Data Base Values

Representative aquifer system hydraulic properties, induced streambed infiltration rates, and contaminant plume properties presented by Heath (1983, p. 13), Davis (1969), Johnson (1967), Morris and Johnson (1967), Davis and DeWiest (1966), Walton (1970), Streltsova-Adams (1978, pp. 360-361), Marsily (1986, pp. 68,79), Rasmussen (1964, pp. 317-325), Polubarinova-Kochina (1962), Domenico (1972, p. 231), Gelhar, et al. (1985), Freeze and Cherry (1979, p. 404-410), and Walton (1988, pp. 19-23, 58) are listed in this appendix to assist the reader in determining the reasonableness of data bases. Values are given for aquifer horizontal hydraulic conductivity, aquitard vertical hydraulic conductivity, aquifer ratio of vertical and horizontal hydraulic conductivities, aquifer specific yield and artesian storativity, aquifer porosity, fractured rock hydraulic properties, and contaminant plume properties and dimensions.

Table F.1   Representative Horizontal Hydraulic Conductivity Values

Rock	Horizontal Hydraulic Conductivity (gpd/sq ft)
Gravel	$1 \times 10^3 - 3 \times 10^4$
Basalt	$1 \times 10^{-6} - 2 \times 10^4$
Limestone	$2 \times 10^{-2} - 2 \times 10^4$
Sand and gravel	$2 \times 10^2 - 5 \times 10^3$
Sand	$1 \times 10^2 - 3 \times 10^3$
Sand, quick	$50 - 8 \times 10^3$
Sand, dune	$1 \times 10^2 - 3 \times 10^2$
Peat, little decomposed	80–300
Peat, moderately decomposed	8–40
Peat, young sphagum	8–80
Peat, old sphagum	6–8
Sandstone	$1 \times 10^{-1} - 50$
Loess	$2 \times 10^{-3} - 20$
Clay	$2 \times 10^{-4} - 2$
Till	$5 \times 10^{-4} - 1$
Shale	$1 \times 10^{-5} - 1 \times 10^{-1}$
Quartzite	$4 \times 10^{-3} - 8$
Greenstone	$1 \times 10^{-1} - 14$
Rhyolite	1–20
Schist	$1 \times 10^{-2} - 2$
Coal	$1 - 1 \times 10^3$

Table F.2   Representative Aquitard Vertical Hydraulic Conductivity

Rock	Aquitard Vertical Hydraulic Conductivity (gpd/sq ft)
Sand, gravel, and clay	$1 \times 10^{-1} - 1 \times 10^0$
Clay, sand, and gravel	$1 \times 10^{-2} - 6 \times 10^{-2}$
Clay	$5 \times 10^{-4} - 1 \times 10^{-2}$
Shale	$1 \times 10^{-7} - 1 \times 10^{-3}$

Table F.3   Representative Stratification Ratios

Degree of Stratification	P_V/P_H Ratio
Low	1/2
Medium	1/10
High	1/100
Very High	1/1000

Table F.4   Representative Specific Yield Values

Rock	Specific Yield (Dimensionless)
Peat	0.30–0.50
Sand, dune	0.30–0.40
Sand, coarse	0.20–0.35
Sand, gravelly	0.20–0.35
Gravel, fine	0.20–0.35
Gravel, coarse	0.10–0.25
Gravel, medium	0.15–0.25
Loess	0.15–0.35
Sand, medium	0.15–0.30
Sand, fine	0.10–0.30
Igneous, weathered	0.20–0.30
Sandstone	0.10–0.40
Sand and gravel	0.15–0.30
Silt	0.01–0.30
Clay, sandy	0.03–0.20
Clay	0.01–0.20
Volcanic, tuff	0.02–0.35
Siltstone	0.01–0.35
Limestone	0.01–0.25
Till	0.05–0.20

Table F.5   Representative Porosity Values

Rock	Porosity (Dimensionless)
Volcanic, pumice	0.80–0.90
Peat	0.60–0.80
Silt	0.35–0.60
Clay	0.35–0.55
Loess	0.40–0.55
Sand, dune	0.35–0.45
Sand, fine	0.25–0.55
Sand, coarse	0.30–0.45
Gravel, coarse	0.25–0.35
Sand and gravel	0.20–0.35
Till	0.25–0.45
Siltstone	0.25–0.40
Sandstone	0.25–0.50
Volcanic, vesicular	0.10–0.50
Volcanic, tuff	0.10–0.40
Limestone	0.05–0.55
Schist	0.05–0.50
Basalt	0.05–0.35
Shale	0.01–0.10
Volcanic, dense	0.01–0.10
Igneous, dense	0.01–0.05
Salt bed	0.005–0.03

Table F.6   Representative Artesian Storativity Values

Material	Artesian Storativity (Dimensionless)
Clay, plastic	6.2×10^{-3}–7.8×10^{-4}m
Clay, stiff	7.8×10^{-3}–3.9×10^{-4}m
Clay, medium hard	3.9×10^{-4}–2.8×10^{-4}m
Sand, loose	3.1×10^{-4}–1.5×10^{-5}m
Sand, dense	6.2×10^{-5}–3.9×10^{-5}m
Sand and gravel, dense	3.1×10^{-5}–1.5×10^{-5}m
Rock, fissured and jointed	2.1×10^{-5}–1.0×10^{-6}m
Rock, sound	$< 1.0 \times 10^{-6}$m

m=aquifer or aquitard thickness (ft)

Table F.7 Representative Induced Streambed Infiltration Rates

Location	Induced Streambed Infiltration Rate (gpd/acre/ft)	Temperature of Surface Water (°F)
Mad River–Springfield, OH	1.0×10^6	39
Sandy Creek–Canton, OH	7.2×10^5	82
Mississippi River–St. Louis, IL	3.1×10^5	54
White River–Anderson, IND	2.2×10^5	69
Miami River–Cincinnati, OH	1.7×10^5	35
Mississippi River–St. Louis, IL	9.1×10^4	33
White River–Anderson, IND	4.0×10^4	38
Mississippi River–St. Louis, IL	3.5×10^4	83

Table F.8 Representative Fractured Rock Hydraulic Property Values

Property	Value
Fracture horizontal hydraulic conductivity	0.01–100 gpd/sq ft
Matrix rock vertical hydraulic conductivity	$1 \times 10^{-6} - 1 \times 10^{-3}$ gpd/sq ft
Fracture porosity (percent of rock thickness)	0.1–1.0
Block porosity (percent of rock thickness)	1.0–30
Fracture width	0.001–1.0 in
Fracture spacing	0.1–10 ft
Block storativity	10^{-6} multiplied by rock thickness
Fracture storativity	$1/10 - 1/100$ of block storativity

Table F.9   Representative Contaminant Plume Property and
Dimension Values

Property or Dimension	Value
Longitudinal dispersivity	0.1 multiplied by plume length up to 300 ft
Transverse dispersivity	$1/5$ to $1/10$ of longitudinal dispersivity
Retardation factor	
Heavy metals	50
Cations	5
Anions	1
Synthetic organic chemicals	1 to 2
Mobility	
Low	when K_d is between 50 and 10^6 mL/g and R_d is between 500 and 5×10^6
Medium	when K_d is between 0.5 and 50 mL/g and R_d is between 5 and 500
High	when K_d is between 10^{-4} and 0.5 mL/g and R_d is between 1 and 5
Plume (elliptical)	
Width	100 to 2000 ft
Length	200 to 10000 ft
Depth	25 to 200 ft
Hydraulic gradient	0.5 to 50 ft/mi
Aquifer bulk mass density	0.7 to 3.2 g/cm^3

Appendix G
Program Verification

Programs WELFUN, WELFLO, CONMIG, and GWGRAF have been checked against published tables of well functions, drawdowns, and concentrations. Selected data bases and associated exact well function, drawdown, and concentration values are presented in this appendix for user verification of program operation.

Table G.1   WELFUN Well FunctionValues

u	W(u)
1E-15	33.96
1E-10	22.45
1E-4	8.633
1E-2	4.038
0.1	1.823
1	0.2194
5	1.148E-3
9	1.245E-5

u	r/B	W(u,r/B)
5.0E-2	1.5	0.4261
7.0E-1	1.5	0.2268
4	1.5	0.0034
5.0E-4	0.7	1.3210
7.0E-2	0.7	1.2438
7	0.7	0.0001
5.0E-6	0.01	9.4413
2.0E-3	0.01	5.6251
1	0.01	0.2144

u_A	BETA	$W(u_A, BETA)$
7.14E-2	4	4.78E-2
4.17E-1	4	4.76E-2
2.5	4	9.33E-3
2.5E-3	0.2	1.08
4.17E-1	0.2	.483
2.5	0.2	2.14E-2
1.25E-4	0.001	5.62
2.5E-2	0.001	2.97
7.14E-1	0.001	.358

u_B	BETA	$W(u_B, BETA)$
4.17E-3	2	4.91
1.25E-2	2	3.82
7.14	2	.178
7.14E-3	0.6	4.37
2.5E-2	0.6	3.14
12.5	0.6	.524
1.25E-3	0.01	6.11
7.14E-3	0.01	4.58
4.17E-1	0.01	3.49

u_f	r/m_b	b	c	$W_{fp}(u_f, r/m_b, b, c)$
1.0E-4	10	1E-5	1E-4	6.4123
1.0E-2	10	1E-5	1E-4	3.0495
1.0E-1	10	1E-5	1E-4	1.5956
1	10	1E-5	1E-4	0.3925

u = 1.0E-3
P_H = 500
P_V = 50
r = 34
m = 50
L = 50
d = 24
L_o = 15
d_o = 13
$W(u, rP_V/P'_H.5/m, L/m, d/m, L_o/m, d_o/m)$ = -1.1457

u = 1.0E-4
PH = 1500
PV = 70
r = 27
m = 50
L = 50
d = 16
Lo = 40
do = 35
$W(u, rP_V/P_H.5/m, L/m, d/m, L_o/m, d_o/m)$ = 1.2644

Table G.2 WELFLO Drawdown Values

Nonleaky artesian aquifer infinite in areal extent, one production well discharging at 500 gpm, aquifer transmissivity is 50000 gpd/ft, and aquifer storativity is 0.0004. Fully penetrating wells.

Time = 100 days

Distance(ft)	Drawdown(ft)
500	11.02
1000	9.43
2000	7.84

Distance = 500 ft

Time(days)	Drawdown(ft)
1	5.75
10	8.38
100	11.02

Leaky artesian aquifer infinite in areal extent, one production well discharging at 100 gpm, aquifer transmissivity is 20000 gpd/ft, aquifer storativity is 0.0003, aquitard thickness is 20 ft, and aquitard vertical hydraulic conductivity is 0.010 gpd/sq ft. Fully penetrating wells.

Time = 30 days

Distance(ft)	Drawdown(ft)
200	4.09
400	3.30
800	2.52

Distance = 200 ft

Time(days)	Drawdown(ft)
3	3.86
30	4.09

Fractured rock artesian aquifer infinite in areal extent, one production well discharging at 400 gpm, fissure horizontal hydraulic conductivity is 50 gpd/sq ft, block vertical hydraulic conductivity is 0.017 gpd/sq ft, storativity of fissured portion of aquifer is 0.00005, storativity of block portion of aquifer is 0.0004, fractured rock aquifer thickness is 600 ft, and half dimension of average block unit is 100 ft. Fully penetrating wells.

Time = 400 days

Distance(ft)	Drawdown(ft)
500	16.03
1000	13.91
2000	11.79

Distance = 500 ft

Time(days)	Drawdown(ft)
4	9.16
40	12.53
400	16.03

Table G.3 CONMIG Concentration Values

One continuous contaminant point source in an aquifer with a thickness of 20 ft, Aquifer actual and effective porosities are 0.3 and 0.2, respectively. Aquifer longitudinal and transverse dispersivities are 50 and 5 ft, respectively for a time interval of 500 days and 100 and 10 ft, respectively for a time interval of 1000 days. The seepage velocity is 1 ft/day. The continuous point source injection rate is 144 gpd. The point source solute concentration is 200 mg/L. Concentrations given below are for monitor wells along the x-axis of the ellipse contaminant plume. There is no adsorption or radioactive decay.

Time = 1000 days

Distance From Point Source(ft)	Concentration(mg/L)
200	5.34
400	3.69
600	2.86

Distance From Point Source = 200 ft

Time(days)	Concentration(mg/L)
500	7.38
1000	5.34

One slug contaminant point source in an aquifer with a thickness of 30 ft. Aquifer actual and effective porosities are 0.25 and 0.18, respectively. Aquifer longitudinal and transverse dispersivities are 60 and 6 ft, respectively for a time interval of 600 days and 150 and 15 ft, respectively for a time interval of 1500 days. The seepage velocity is 1 ft/day. The slug point source contaminant load is 10 pounds. Concentrations given below are for monitor wells along the x-axis of the ellipse contaminant plume, There is no adsorption or radioactive decay.

Time = 600 days

Distance From Point Source(ft)	Concentration(mg/L)
200	0.07
400	0.16
600	0.21

Distance From Point Source = 600 ft

Time(days)	Concentration(mg/L)
600	0.21
1500	0.01

Appendix H
Example Program Input/Output Displays

Four example input/output displays to guide the user through WELFUN, WELFLO, CONMIG, and GWGRAF program operation are presented in this appendix. Data bases constitute input and computation, results constitute output. The information presented was displayed on paper during program operation by a printer.

Example H.1 is for a partially penetrating observation well in a water table aquifer. Example H.2 is for a leaky artesian aquifer with 2 fully penetrating production wells without wellbore storage. The aquifer has no boundaries. Example H.3 is for one continuous point source with known injection rate and solute concentration. There is no sorption or radioactive decay. Example H.4 is for scattered water level data.

Example H.1   WELFUN input/output

DATA BASE:

uB= 1.2000E+00
Beta= 2.4000E-01

COMPUTATION RESULTS:

W(uB,Beta)= 1.1229E+00

PARTIAL PENETRATION DATA BASE:

Aquifer horiz. hydraulic conductivity (gpd/sq ft)= 5.0000E+02
Aquifer vert. hydraulic conductivity (gpd/sq ft)= 5.0000E+01
Radial distance to well in ft= 2.4000E+01
Aquifer thickness in ft= 3.5000E+01
Dist. from aquifer top to bottom of prod. well (ft)= 3.5000E+01
Dist. from aquifer top to top of prod. well screen(ft) 1.7000E+01
Dist. from aquifer top to bottom of obs. well (ft)= 3.3000E+01
Dist. from aquifer top to top of obs. well screen(ft)= 3.1000E+01

COMPUTATION RESULTS:

W(uB,Beta,rPv/Ph^.5/m,L,/m,d/m,Lo/m,do/m)= 1.2697E+00
W(u,rPv/Ph^.5/m,L/m,d/m,Lo/m,do/m)= 1.4685E-01

Example H.2   WELFLO input/output

GENERAL DATA BASE:

Number of simulation periods for which drawdown
or recovery is to be calculated 2
Simulation period number= 1
Duration of simulation period in days=   10.000
Simulation period number= 2
Duration of simulation period in days=   200.000
Number of grid columns= 8
Number of grid rows= 8
Grid spacing in ft=   500.00
X-coordinate of upper-left grid node in ft=   0.00
Y-coordinate of upper-left grid node in ft=   0.00
Simulation period number= 1
Number of production, injection, and image wells
active during simulation period= 2
Well number= 1
X-coordinate of well in ft=   2000.00
Y-coordinate of well in ft=   2000.00
Well discharge in gpm=   250.00
Duration of pump operation during simulation period
in days=   10.000
Well radius in ft=   0.30
Simulation period number= 1
Number of production, injection, and image wells
active during simulation period= 2
Well number= 2
X-coordinate of well in ft=   1000.00
Y-coordinate of well in ft=   1000.00
Well discharge in gpm=   150.00
Duration of pump operation during simulation period
in days=   10.000
Well radius in ft=   0.40
Simulation period number= 2
Number of production, injection, and image wells
active during simulation period= 2
Well number= 1
X-coordinate of well in ft=   2000.00
Y-coordinate of well in ft=   2000.00

Well discharge in gpm= 250.00
Duration of pump operation during simulation period
in days= 200.000
Well radius in ft= 0.30
Simulation period number= 2
Number of production, injection, and image wells
active during simulation period= 2
Well number= 2
X-coordinate of well in ft= 1000.00
Y-coordinate of well in ft= 1000.00
Well discharge in gpm= 150.00
Duration of pump operation during simulation period
in days= 200.000
Well radius in ft= 0.40
Number of observation wells for which time-
drawdown tables are desired 1
Observation well number= 1
I-coordinate of observation well= 6
J-coordinate of observation well= 6
Aquifer transmissivity in gpd/ft= 20000.00
Aquifer storativity as a decimal= 0.000400
Aquitard thickness in ft= 100.00
Aquitard vert. hydr. conduct. in gpd/sq ft= 0.001

NODAL COMPUTATION RESULTS:

SIMULATION PERIOD DURATION IN DAYS: 10.000

VALUES OF DRAWDOWN OR RECOVERY (FT) AT NODES:

J-ROW				I-COLUMN				
	1	2	3	4	5	6	7	8
1	7.91	8.66	9.16	9.20	8.88	8.40	7.81	7.19
2	8.66	9.90	10.96	10.70	10.06	9.31	8.55	7.74
3	9.16	10.96	23.87	12.28	11.41	10.40	9.26	8.23
4	9.20	10.70	12.28	13.00	13.21	11.62	9.88	8.53
5	8.88	10.06	11.41	13.21	34.06	12.41	10.06	8.59
6	8.40	9.31	10.40	11.62	12.41	11.14	9.55	8.31
7	7.81	8.55	9.26	9.88	10.06	9.55	8.68	7.82
8	7.19	7.74	8.23	8.53	8.59	8.31	7.82	7.20

NODAL COMPUTATION RESULTS:

SIMULATION PERIOD DURATION IN DAYS: 200.000

VALUES OF DRAWDOWN OR RECOVERY (FT) AT NODES:

J-ROW				I-COLUMN				
	1	2	3	4	5	6	7	8
1	13.48	14.24	14.75	14.79	14.47	13.97	13.37	12.73
2	14.24	15.49	16.55	16.33	15.69	14.95	14.12	13.29
3	14.75	16.55	29.50	17.92	17.05	16.03	14.86	13.79
4	14.79	16.33	17.92	18.63	18.84	17.25	15.49	14.13
5	14.47	15.69	17.05	18.84	39.69	18.02	15.67	14.19
6	13.97	14.95	16.03	17.25	18.02	16.75	15.16	13.90
7	13.37	14.12	14.86	15.49	15.67	15.16	14.27	13.37
8	12.73	13.29	13.79	14.13	14.19	13.90	13.37	12.73

OBSERVATION WELL COMPUTATION RESULTS WITH FULL WELL
PENETRATION AND WITHOUT WELLBORE STORAGE:

TIME-DRAWDOWN OR RECOVERY TABLE

OBSERVATION WELL NUMBER: 1

TIME(DAYS)	DRAWDOWN OR RECOVERY(FT)
10.000	11.14
200.000	16.75

Example H.3   CONMIG input/output

DATA BASE:

Number of simulation periods for which contaminant
concentration distribution is to be calculated 1

Simulation period number= 1
Simulation period duration in days= 1000.00
Number of grid columns= 8
Number of grid rows= 8
Grid spacing in ft=   100.00
X-coordinate of upper-left grid node in ft=   0.00
Y-coordinate of upper-left grid node in ft=   0.00
Aquifer actual porosity as a decimal= 0.300
Aquifer effective porosity as a decimal= 0.200
Simulation period number= 1
Aquifer thickness in ft= 24.00
Aquifer longitudinal dispersivity in ft=100.00
Aquifer transverse dispersivity in ft= 10.00
Seepage velocity in ft/day=   1.00
Number of point sources= 1
Simulation period number= 1
Point source number 1
X-coordinate of point source in ft=   100.00
Y-coordinate of point source in ft=   300.00
Continuous point source injection rate in gpd=   1000.00
Point source solute concentration in mg/l=   345.000
Time after continuous contaminant injection
started in days= 1000.00
Number of monitor wells for which time-
concentration tables are desired= 1
Monitor well number= 1
I-coordinate of monitor well= 4
J-coordinate of monitor well= 4

NODAL COMPUTATION RESULTS:

SIMULATION PERIOD DURATION IN DAYS: 1000.00

VALUES OF CONTAMINANT CONCENTRATION (MG/L) AT NODES:

J-ROW				I-COLUMN				
	1	2	3	4	5	6	7	
1	0.08	0.13	0.21	0.31	0.42	0.54	0.65	
2	0.66	1.15	1.79	2.54	3.28	3.92	4.36	
3	4.77	8.72	12.96	16.09	18.66	18.92	18.27	1
4	26.73	203.30	72.66	53.29	43.60	36.84	33.29	2
5	4.77	8.72	12.96	16.09	18.66	18.92	18.27	1
6	0.66	1.15	1.79	2.54	3.28	3.92	4.36	
7	0.08	0.13	0.21	0.31	0.42	0.54	0.65	
8	0.01	0.01	0.02	0.03	0.04	0.05	0.07	

MONITOR WELL COMPUTATION RESULTS:

TIME-CONCENTRATION TABLE

MONITOR WELL NUMBER: 1

TIME(DAYS)	CONCENTRATION(MG/L)
1000.000	53.29

Example H.4   GWGRAF input/output

DATA BASE:

Number of data points= 8
Data point number= 1
Z-value of data point in ft or mg/l=   703.00
X-coordinate of data point in ft=   750.00
Y-coordinate of data point in ft=   750.00
Data point number= 2
Z-value of data point in ft or mg/l=   696.00
X-coordinate of data point in ft=   1700.00
Y-coordinate of data point in ft=   700.00
Data point number= 3
Z-value of data point in ft or mg/l=   689.00
X-coordinate of data point in ft=   2400.00
Y-coordinate of data point in ft=   800.00
Data point number= 4
Z-value of data point in ft or mg/l=   697.00
X-coordinate of data point in ft=   1250.00
Y-coordinate of data point in ft=   1300.00
Data point number= 5
Z-value of data point in ft or mg/l=   688.00
X-coordinate of data point in ft=   1600.00
Y-coordinate of data point in ft=   1700.00
Data point number= 6
Z-value of data point in ft or mg/l=   684.00
X-coordinate of data point in ft=   2200.00
Y-coordinate of data point in ft=   1600.00
Data point number= 7
Z-value of data point in ft or mg/l=   689.00
X-coordinate of data point in ft=   800.00
Y-coordinate of data point in ft=   2400.00
Data point number= 8
Z-value of data point in ft or mg/l=   678.00
X-coordinate of data point in ft=   2300.00
Y-coordinate of data point in ft=   2400.00
Number of grid columns= 5
Number of grid rows= 5
Grid spacing in ft=   500.00
X-coordinate of upper-left grid node in ft=   500.00
Y-coordinate of upper-left grid node in ft=   500.00

NODAL COMPUTATION RESULTS:

VALUES OF Z IN FT OR MG/L AT NODES:

J-ROW				I-COLUMN			
	1	2	3	4	5	6	7
1	702.70	701.90	696.16	694.08	689.35		
2	702.27	700.00	696.30	691.77	688.90		
3	697.12	696.27	690.46	684.50	684.30		
4	689.81	689.76	688.17	683.51	680.85		
5	689.04	688.99	685.97	678.76	678.13		

References

Abramowitz, M., and I.A. Stegun, Eds. 1964. Handbook of Mathematical Functions with Formulas, Graphs, and Mathematical Tables. U.S. Dept. of Commerce, National Bureau of Standards. Applied Mathematics Series, Vol. 55.

Anderson, M.P. 1984. Movement of Contaminants in Groundwater: Groundwater Transport-Advection and Dispersion. In Groundwater Contamination. National Academy Press.

Bear, J. 1979. Hydraulics of Groundwater. McGraw-Hill Book Company.

Bear, J., and A. Verruijt. 1987. Modeling Groundwater Flow and Pollution. D. Reidel Publishing Co.

Boulton, N.S., and T.D. Streltsova. 1977. Unsteady Flow to a Pumped Well in a Fissured Water-Bearing Formation. Journal of Hydrology. Vol. 35.

Bourke, P.D. 1987. A Contouring Subroutine. BYTE Publications, Inc. June.

Case, C.M., and J.C. Addiego. 1977. Note on a Series Representation of the Leaky Aquifer Well Function. Journal of Hydrology. Vol. 32.

Cherry, J.A., R.W. Gillham, and J.F. Barker. 1984. Contaminants in Groundwater: Chemical Processes. In Groundwater Contamination. National Academy Press.

Clark, D. 1987. Microcomputer Programs for Groundwater Studies. Elsevier Science Publishing Co., Inc.

Cleary, R.W., and M.J. Ungs. 1978. Groundwater Pollution

and Hydrology, Mathematical Models and Computer Programs. Water Resources Program, Princeton University Report 78-WR-15.

Codell, R.B., K.T.Key, and G.Whelan. 1981. A Collection of Mathematical Models for Dispersion in Surface Water and Groundwater. Division of Engineering, Office of Nuclear Reactor Regulation, U.S. Nuclear Regulatory Commission. NUREG-0868.

Cobb, P.M., C.D. Mc Elwee, and M.A. Butt. 1982. An Automated Numerical Evaluation of Leaky Aquifer Pumping Test Data, etc. Groundwater Series 6. Kansas Geological Survey.

Cooper, H.H.,Jr. and C.E. Jacob. 1946. A Generalized Graphical Method for Evaluating Formation Constants and Summarizing Well Field History. Transactions of the American Geophysical Union.

Dansby, D.A. 1987. Graphical Well Analysis Package – A Graphical, Computer-Assisted, Curve Matching Approach to Well Test Analysis. In Proceedings of the Solving Ground Water Problems with Models Conference and Exposition. February 10-12, 1987, Denver.

Da Prat, G. 1981. Well Test Analysis for Naturally-Fractured Reservoirs. PhD Dissertation. Department of Petroleum Engineering, Stanford University.

Davis, J.C. 1986. Statistics and Data Analysis in Geology. John Wiley & Sons, Inc.

Davis, S.N., and R.J.M. DeWiest. 1966. Hydrogeology. John Wiley & Sons, Inc.

Davis, S.N. 1969. Porosity and Permeability in Natural Materials. In Flow through Porous Media. R.J.M. DeWiest, Ed. Academic Press, Inc.

Domenico, P.A. 1972. Concepts and Models in Groundwater Hydrology. McGraw-Hill Book Company.

Dougherty, D.E., and D.K. Babu. 1984. Flow to a Partially Penetrating Well in a Double-Porosity Reservoir. Water Resources Research. Vol.20, No. 8.

Driscoll, F.G. 1986. Groundwater and Wells. 2nd ed. Johnson Division, Signal Environmental Systems, Inc., New Brighton, MN.

Fenske, P.R. 1984. Unsteady Drawdown in the Presence of a Linear Discontinuity. In Groundwater Hydraulics. American Geophysical Union Monograph 9.

Ferris, J.G., D.B. Knowles, R.H. Brown, and R.W. Stallman. 1962. Theory of Aquifer Tests. U.S. Geological Survey. Water Supply Paper 1536-E.

Fowler, J. 1984. IBM PC/XT Graphics Book. Prentice-Hall, Inc.

Freeze, R.A., and J.A. Cherry. 1979. Groundwater. Prentice-Hall, Inc.

Gelhar, L.W., A. Mantoglou, C. Welty, and K.R. Rehfeldt. 1985. A Review of Field Scale Physical Solute Transport Processes in Saturated and Unsaturated Media. Electric Power Research Institute. Palo Alto, CA. Report No. EPRI EA-4190.

Goldstein, L.J. 1984. Advanced BASIC and Beyond for the IBM PC. Robert J. Brady Co.

Hantush, M.S., and C.E. Jacob. 1955. Non-Steady Radial Flow in an Infinite Leaky Aquifer. Transactions of the American Geophysical Union. Vol. 36, No. 1.

Hantush, M.S. 1961. Drawdown Around a Partially Penetrating Well. Hydraulics Division, Proceedings of the American Society of Civil Engineers.

Hantush, M.S. 1964. Hydraulics of Wells. In Advances in Hydroscience. Vol. 1. Academic Press, Inc.

Harris, J.M., and M.L.Scofield. 1983. IBM PC Conversion Handbook of BASIC. Prentice-Hall, Inc.

Hearn, D. and M.P. Baker. 1983. Computer Graphics for the IBM Personal Computer. Prentice-Hall, Inc.

Heath, R.C. 1983. Basic Ground-Water Hydrology. U.S Geological Survey Water-Supply Paper 2220.

Hunt, B. 1977. Calculation of the Leaky Aquifer Function. Journal of Hydrology. Vol. 33.

Hunt, B. 1983. Mathematical Analysis of Groundwater Resources. Butterworth & Co., Ltd.

Huisman, L., and T.N. Olsthoorn. 1983. Artificial Groundwater Recharge. Pitman Advanced Publishing Program, Boston, MA.

Huyakorn, P.S. and G.F. Pinder. 1983. Computational Methods in Subsurface Flow. Academic Press, Inc.

Jacob, C.E. 1944. Notes on Determining Permeability by Pumping Tests Under Water Table Conditions. U.S. Geological Survey. Mimeographed Report.

Jacob, C.E. 1946. Drawdown Test to Determine Effective Radius of Artesian Well. Proceedings of the American Society of Civil Engineers. Vol. 79, No. 5.

Javandel, I., C. Doughty, and C.F. Tsang. 1984. Groundwater Transport: Handbook of Mathematical Models. American Geophysical Union Water Resources Monograph Series 10.

Jones, T.A., D.E. Hamilton, and C.R.Johnson. 1986. Contouring Geologic Surfaces with the Computer. Van Nostrand Reinhold Company.

Johnson, A.I. 1967. Specific Yield-Compilation of Specific Yields for Various Materials. U.S. Geological Survey. Water Supply Paper 1662-D.

Kinzelbach, W. 1986. Groundwater Modelling – An Introduction with Sample Programs in BASIC. Elsevier Science Publishers, Amsterdam.

Korites, B.J. 1982. Data Plotting Software for Micros. Kern Publications.

Krylov, V.I. 1962. Approximate Calculation of Integrals. Macmillan Publishing Co., Inc.

Lancaster, P., and K. Salkauskas. 1986. Curve and Surface Fitting – An Introduction. Academic Press, Inc.

Marsily, Ghislain de. 1986. Quantitative Hydrogeology. Academic Press, Inc.

Miller, K.S. 1957. Engineering Mathematics. Rinehart and Co., Inc.

Miller, A.R. 1981. BASIC Programs for Scientists and Engineers. SYBEX Inc., Oakland, CA.

Miller, C.T., and W.J. Weber. 1984. Modeling Organic Contaminant Partitioning in Ground-Water Systems. Ground Water. Vol. 22, No. 5.

Moench, A.F., and A. Ogata. 1981. A Numerical Inversion of the Laplace Transform Solution to Radial Dispersion in a Porous Medium. Water Resources Research. Vol. 17, No. 1.

Moench, A.F. 1984. Double-Porosity Models for a Fissured Groundwater Reservoir with Fracture Skin. Water Resources Research. Vol. 20.

Moench, A.F., and A. Ogata. 1984. Analysis of Constant Discharge Wells by Numerical Inversion of Laplace Transform Solutions. In Groundwater Hydraulics. American Geophysical Union. Water Resources Monograph 9.

Morris, D.A., and A.I. Johnson. 1967. Summary of Hydrologic and Physical Properties of Rock and Soil Materials, as Analyzed by the Hydrologic Laboratory of the U.S. Geological Survey 1948–1960. U.S. Geological Survey. Water Supply Paper 1839-D.

Muskat, M. 1937. The Flow of Homogeneous Fluids through Porous Medium. McGraw-Hill Book Company.

Neuman, S.P., and Witherspoon, P.A. 1969. Applicability of Current Theories of Flow in Leaky Aquifers. Water Resources Research. Vol. 5, No. 4.

Neuman, S.P. 1975a. Analysis of Pumping Test Data from Anisotropic Unconfined Aquifers Considering Delayed Gravity Response. Water Resources Research. Vol. 11, No. 2.

Neuman, S.P. 1975b. A Computer Program to Calculate Drawdown in an Anisotropic Unconfined Aquifer with a Partially Penetrating Well. Unpublished Manuscript. Department of Hydrology and Water Resources, University of Arizona.

Olea, R.A. 1975. Optimum Mapping Techniques Using Regionalized Theory. Kansas Geological Survey Series on Spatial Analysis. No. 2.

Polubarinova-Kochina, P.YA. 1962. Theory of Ground Water Movement. Princeton University Press.

Poole, L., M. Borchers, and K. Koessel. 1981. Some Common BASIC Programs. Osborne/McGraw-Hill, Inc.

Press, W.H., B.P. Flannery, S.A. Teukolsky, and W.T. Vetterling. 1986. Numerical Recipes–The Art of Scientific Computing. Cambridge University Press.

Prickett, Thomas A. 1981. Oral communication. Champaign, IL.

Rasmussen, W.C. 1964. Permeability and Storage of Heterogeneous Aquifers in the U.S. International Association of Scientific Hydrology. Publication 64.

Raudkivi, A.J., and R.A. Callander. 1976. Analysis of Groundwater Flow. John Wiley & Sons, Inc.

Reed, J.E. 1980. Type Curves for Selected Problems of Flow to Wells in Confined Aquifers. Techniques of Water-Resources Investigations of the U.S. Geological Survey. Book 3, Chapter B3.

Rushton, K.R., and S.C. Redshaw. 1979. Seepage and Groundwater Flow. John Wiley & Sons, Ltd.

Sandberg, R., R.B. Scheiback, D. Koch, and T.A. Prickett. 1981. Selected Hand-Held Calculator Codes for the Evaluation of the Probable Cumulative Hydrologic Impacts of Mining. U.S. Dept. of the Interior, Office of Surface Mining. H–D3004/030–81–1029F.

Sandler, C. 1984. Desktop Graphics for the IBM PC. Creative Computing Press.

Sauveplane, C.M. 1984. On the Use of Approximate Analytical Inversion of Laplace Transform for Radial Flow Problems. In International Groundwater Symposium on Groundwater Resources Utilization and Contaminant Hydrogeology. Vol. 1. Atomic Energy of Canada Ltd.

Schapery, R.A. 1961. Approximate Methods of Transform Inversion for Viscoelastic Stress Analysis. 4th Proceedings of the U.S. National Congress of Applied Mechanics.

Shad/Chen/Frank. 1964. Tables of Zeros and Gaussian Weights of Certain Associated Laguerre Polynomials and the Related Generalized Hermite Polynomials. IBM Technical Report TR00, 1100.

Simons, S.L. November, 1983. Make Fast and Simple Contour Plots on a Microcomputer. BYTE Publications, Inc.

Stallman, R.W. 1963. Type Curves for the Solution of Single-Boundary Problems. In Bentall, R., Compiler. Shortcuts and Special Problems in Aquifer Tests. U.S. Geological Survey. Water Supply Paper 1545-C.

Stehfest, H. 1970. Algorithm 368. Numerical Inversion of Laplace Transforms. Commun. ACM. Vol. 13, No. 1.

Streltsova-Adams, T.D. 1978. Well Hydraulics in Heterogeneous Aquifer Formations. In Advances in Hydroscience. Ven Te Chow, Ed. Vol. 11. Academic Press, Inc.

Streltsova, T.D. 1988. Well Testing in Heterogeneous Formations. John Wiley & Sons, Inc.

Theis, C.V. 1935. The Relation Between the Lowering of the Piezometric Surface and the Rate and Duration of Discharge of a Well Using Ground-Water Storage. Transactions of the American Geophysical Union. Vol. 16, Part 2.

Walton, W.C. 1970. Groundwater Resource Evaluation. McGraw-Hill Book Company.

Walton, W.C. 1984a. Analytical Groundwater Modeling with Programmable Calculators and Hand-Held Computers. In Groundwater Hydraulics. American Geophysical Union. Water Resources Monograph 9.

Walton, W.C. 1984b. Thirty-five BASIC Groundwater Programs for Desktop Microcomputers. International Ground Water Modeling Center, Holcomb Research Institute, Butler University WALTON84-BASIC.

Walton, W.C. 1987. Groundwater Pumping Tests. Lewis Publishers, Inc.

Walton, W.C. 1988. 3rd ed. Practical Aspects of Groundwater Modeling. National Water Well Association.

Watson, G.N. 1958. A Treatise on the Theory of Bessel Functions. Cambridge University Press.

Weinman, D.G., and B.L.Kurshan. 1985. IBM PC BASIC for Scientists and Engineers. Reston Publishing Company.

Wilkes, M.V. 1966. A Short Introduction to Numerical Analysis. Cambridge University Press.

Wilson, J.L., and P.J. Miller. 1978. Two-Dimensional Plume in Uniform Groundwater Flow. Journal of the Hydrology Division, American Society of Civil Engineers. Vol. 104, No. HY 4.

Yeh, G.T. 1981. AT123D: Analytical Transient One-, Two-, and Three-Dimensional Simulation of Waste Transport in the Aquifer System. IGWMC MARS #6120. ORNL-5602. Oak Ridge National Laboratory. Environmental Sciences Division Publication No. 1439.

Index